T0276211

# CAMBRIDGE LIBRARY COLLECTION

*Books of enduring scholarly value*

## Physical Sciences

From ancient times, humans have tried to understand the workings of the world around them. The roots of modern physical science go back to the very earliest mechanical devices such as levers and rollers, the mixing of paints and dyes, and the importance of the heavenly bodies in early religious observance and navigation. The physical sciences as we know them today began to emerge as independent academic subjects during the early modern period, in the work of Newton and other 'natural philosophers', and numerous sub-disciplines developed during the centuries that followed. This part of the Cambridge Library Collection is devoted to landmark publications in this area which will be of interest to historians of science concerned with individual scientists, particular discoveries, and advances in scientific method, or with the establishment and development of scientific institutions around the world.

## An Experimental Inquiry into the Nature and Propagation of Heat

The Scottish mathematician and natural philosopher Sir John Leslie (1766–1832) had set out at the end of the eighteenth century to explore the nature of heat radiation, which he felt was a 'dubious and neglected' area of physics. Leslie's inquiry, published in 1804, details his many experiments, notably the use of two self-devised instruments: Leslie's cube and his differential thermometer. Establishing several basic laws of heat radiation and rejuvenating the debate about the physical composition of heat, Leslie's work gained him the Rumford medal of the Royal Society in 1805. Nevertheless, the same publication jeopardised his chances of obtaining an academic position at Edinburgh. A single, allegedly atheistic endnote, supporting David Hume's views on causation, prompted protests by the local clergymen when his candidature for the chair of mathematics was under consideration. Leslie secured the professorship, however, and remained with the university until his death.

# An Experimental Inquiry into the Nature and Propagation of Heat

JOHN LESLIE

**CAMBRIDGE**
**UNIVERSITY PRESS**

University Printing House, Cambridge, CB2 8BS, United Kingdom

Cambridge University Press is part of the University of Cambridge.
It furthers the University's mission by disseminating knowledge in the pursuit of education, learning and research at the highest international levels of excellence.

www.cambridge.org
Information on this title: www.cambridge.org/9781108080170



This edition first published 1804
This digitally printed version 2015

ISBN 978-1-108-08017-0 Paperback



The original edition of this book contains a number of oversize plates which it has not been possible to reproduce to scale in this edition.
They can be found online at www.cambridge.org/9781108080170

AN

*EXPERIMENTAL*

# INQUIRY

INTO

*THE NATURE,*

AND

*PROPAGATION,*

OF

# HEAT.

BY JOHN LESLIE.

LONDON:
PRINTED FOR J. MAWMAN, NO. 22 POULTRY;
SOLD ALSO BY BELL AND BRADFUTE,
EDINBURGH.

1804.

T. Gillet Printer, Salisbury Square.

TO

# THOMAS WEDGWOOD, ESQ.

IN presenting this volume to my best friend, I feel the satisfaction of having, at least in part, discharged an obligation which springs from the sense of duty. To you principally it owes its birth. By your counsels and generous encouragement, I have obeyed an early impulse, resolved to cherish the spirit of independence, and devote my life to the attractive pursuits of philosophy. I prize it as the highest distinction, to have so long shared your affectionate esteem. At a more fortunate period, we lived in the closest intimacy, united by kindred sentiments and actuated by similar views. Ever shall I deplore the cruel interruption of your eager career. You have had to maintain a continued struggle with that languid state of health which is but too often the companion of genius and exquisite sensibility. Happily your mind, superior to the corrosion of pain, still preserves its native lustre. You take the same interest in all the various concerns of humanity, and anxiously seek to promote whatever may eventually contribute to the improvement of our species.

As the real wants of man are always physical, an intimate acquaint-

ance

ance with the laws of Nature, while it expands his mental energies, has an evident tendency to ameliorate his actual condition. The present inquiry has, therefore, a farther claim to your notice. It will recall, though with a mixture of regret, some pleasing associations. On a subject thus interesting, your correct information, guided by a refined taste and habits of logical precision, eminently qualify you to decide. If this work shall obtain your approbation, I may dismiss all solicitude for its reception with a discerning public.

JOHN LESLIE.

LARGO, FIFESHIRE,
*20th March*, 1804.

# PREFACE.

THE leading facts which gave rise to this publication presented themselves in the spring of 1801. I was transported at the prospect of a new world emerging to view. Without delay I determined to investigate the subject with that close attention which it seemed to demand. I planned a course of experiments, and having procured the necessary apparatus, I retired to my native spot, where I could advantageously push the research in calm seclusion. The summer and autumn were mostly spent in this delightful employment; and I prosecuted my inquiries with ardour, with unremitting diligence, and pro-

 portionate

portionate success. The results gradually converged, and pointed to a few connecting principles, which appeared to coalesce into a regular system. I now set about digesting those ideas, and in the course of the following winter, the composition of the present volume was considerably advanced. The gleam of peace tempted me to indulge a temporary suspension; and I repaired, with the crowd of my countrymen, to the famed capital, where the treasures of art and science are so profusely displayed. In that vortex of pleasure and centre of information, I spent several months very agreeably. But the task which I had undertaken imperiously recalled my thoughts. I returned to London near the close of the year, and having committed the manuscript to the press, I hastened again to my retreat. I enlarged the apparatus, performed some of the experiments anew with different modifications, and carried forward a number of subordinate inquiries calculated to improve

and

and extend the theory. But while I was thus engaged, the execution of the work proceeded with extreme tardiness: it met with repeated interruption, and has experienced such unexpected and provoking delays as had almost exhausted my stock of patience. The charm of novelty is worn off, and I begin to look upon my production with a coolness, I believe, not usual in authors.

To mention these particulars is not altogether superfluous. They will explain the slight variation of tone, and perhaps defect of unity, that may be perceived in the progress of the disquisition. My distance from the press, besides occasioning other inconvenience, has prevented me from bestowing the degree of correction which I was solicitous to attain. Some mistakes of a more important kind may have escaped notice, which I should probably have rectified, if the impression had been more immediately under my command.

The

The object I chiefly proposed, was to discover the nature, and ascertain the properties, of what is termed Radiant Heat. No part of physical science appeared so dark, so dubious and neglected. Reflection had long taught me to consider the communication of heat among insulated bodies as performed only by the medium of the intervening air. This opinion I now put beyond dispute. But my researches went farther, and laid open an assemblage of facts, not more unexpected than curious and important. Viewed in their mutual bearings, they gave some glimpse of the recondite mode of operation, and disclosed the nice conditions that determine such atmospheric agency. Thus directed, I could advance securely in tracing their extensive consequences. But to develope the latent constitution of heat, required much abstruse and elaborate discussion. In this difficult inquiry, I have endeavoured to tread with cautious steps: I have carefully avoided hypotheses, and

where

where actual observation failed, I have sought the guidance of analogical reasoning. To support or illustrate my arguments, I have introduced a variety of original matter, the fruit of prior research, and which, I hope, will be found interesting. The propagation of heat I have examined at great length: I have distinguished the elements that enter into that complex process, and have estimated their separate influence. Furnished with such principles, I could demonstrate in the most convincing manner the theory of the Photometer. I have fully described the construction of that curious instrument, and have given a rapid sketch of the various applications for which it is fitted. But it yet remained to inquire how far the communication of heat is affected by the state or quality of the ambient medium, or to distinguish the relative powers of transmission that belong to the several permanent gases, and are modified by the different mea-

sures

sures of rarefaction. With this arduous investigation, the volume concludes.

It may seem a very material omission, not to have inquired into the cause of the production of heat by friction. Respecting this intricate subject, I have collected some ideas, which, if they were fully matured, will perhaps afford a satisfactory explication ; but having already exceeded the ordinary limits, I have left abundant store in reserve. I purpose, in a subsequent volume, to resume and extend my researches. I shall then trace the practical consequences that flow from the principles now established, and point out in detail their application to the management of artificial heat. This discussion will suggest a number of improvements conducive to the comforts or elegant luxuries of life. I shall next examine the heat derived from the continual irradiation of the sun, and explain its distribution by by the vehicle of atmospheric commerce over the surface of our globe. The circumstances relative to climate deserve particular investigation, and I shall, from theory corrected by observations, deduce the mean temperatures, whether moulded by latitude or elevation. The transition is natural from heat to humidity: I shall, therefore, unfold the mutual affections of air and moisture, and exhibit the results of my hygrometrical inquiries. And by judiciously combining these principles, I may prepare a solid foundation for erecting a system of Meteorology.

I have seldom ventured into the region of conjecture, but have patiently sought to determine facts with accuracy, and have laboured to deduce the consequences by a close train of argument, expressed in concise and forcible language. Experiments, in the present improved state of physics, owe their whole value to the care and precision with which they are

conducted.

conducted. It will scarcely, I presume, be denied, that I have at least searched more deeply than those who have gone before me. I have employed instruments of delicate construction, and have digested and compared the results with scrupulous assiduity. In analysing the principles and estimating their compound effects, I have derived important assistance from the application of mathematical science.

I am very far, however, from thinking the subject yet exhausted. I have spent much time, and incurred no small expence, in this research. But it would be very desirable to repeat the experiments on a larger scale, and with more extensive apparatus. I would invite other inquirers to dispel what obscurity yet remains, and to correct the mistakes which I may have committed. Throughout the whole, I have freely exercised my reason, unawed by authority and uninfluenced by current opinion.

opinion. In the course of investigation, I have found myself compelled to relinquish some preconceived notions; but I have not abandoned them hastily, nor, till after a warm and obstinate defence, I was driven from every post. The conclusions which I have adopted must hence rest on a surer basis. I might therefore bespeak the candid attention of the philosophic reader. I have no desire to shrink from liberal criticism ; but I request my book to be perused and examined with the same temper in which it was written.

## DIRECTIONS FOR THE BOOKBINDER.

| *These Plates to fold out.* | | | *The following not to fold.* | | |
|---|---|---|---|---|---|
| Plate I. to face the title page. | | | Plate VI. | to face page | 424. |
| II. | - | p. 514. | VII. | - | p. 426. |
| III. | - | p. 520. | VIII. | - | p. 468. |
| IV. | - | p. 284. | IX. | - | p. 490. |
| V. | - | p. 436. | | | |

### ERRATA.

Page 344. Experiment LII. *is marked* LI, and this mistake runs through the rest of the volume.

336, line 6 from the bottom, *for* 82, *read* 80.

337, —- 15, *for* .55 and 77, *read* .50 and 80.

17, *read* $\frac{50}{80}\left(\text{Log.}\ {}^{20}_{10} - \text{Log.}\ \frac{100}{90}\right) = 159\frac{1}{2}$.

18, *for* 82, *read* 80.

525, —- 6, *for* casual *read* causal.

AN

# EXPERIMENTAL INQUIRY

INTO THE

# NATURE

AND

# PROPAGATION OF HEAT.

## CHAPTER I.

BEFORE I proceed to relate the experiments which serve as the basis of the reasonings contained in the ensuing Tract, it will be proper to give some idea of the form and construction of the instruments which were employed. With this previous information, it will be more easy to comprehend the subsequent detail, and to judge what degree of credit is due to the accuracy of the experiments themselves. I shall likewise be thence enabled to avoid much repetition, and to spare the patience of my reader, which would be exposed to a severe trial in the perusal of monotonous de-

 scriptions

scriptions of instruments and manipulations. Nothing, surely, is more requisite in physical inquiries, than to mark carefully the circumstances that enter into the process of an experiment, and to exclude, with the utmost solicitude, every thing which, by a foreign influence, might disturb or alter the result. Yet, after all, much must be left to the fidelity and skill of the experimenter; and to attempt, by a prolix recital, to convey distinct and adequate ideas of a combination of objects which would at once, and without effort, forcibly strike the senses, will in most cases prove a task equally fruitless and irksome.

The principal articles of the apparatus were *specula* or reflectors made of tinned-iron. Of these I had several, of different dimensions; from twelve to about fourteen inches in diameter, and with a depth of concavity from 1¾ to near 2½ inches. It cost me no small trouble to obtain what I wanted. I had to make repeated trials before I could find an artist skilful enough to execute the reflectors with any tolerable precision, or who was disposed to listen to my directions. But, by dint of perseverance, my sanguine

guine wishes were at length gratified. The reflectors were hammered out of block-tin, and highly finished, exhibiting an admirable brightness, smoothness, and regularity of surface. Aware that the aberration from the focus in reflection must be very considerable when large segments of hollow spheres are used, I sought to procure the parabolic figure. I formed thin slips of mahogany with great accuracy, into segments of parabolas of different sizes, to serve as gages for the workman;* and with some dexterity, and the frequent changing of the hammers, the reflectors were fashioned to fit those shapes with surprizing exactness. As it was my object to obtain the most powerful reflectors that could be made in this way, I was at pains to procure tin-plates of the largest dimensions, and to have them hammered to the greatest depth of concavity that the metal would bear without being fractured. Exposed to the direct light of the sun, these reflectors collected the rays into a pretty distinct focus, scarcely exceeding half an inch in diameter, so

* See Note I.

that the whole errors of the figure did not occasion a deviation of more than a quarter of an inch on either side. And though the sun's image was therefore one hundred times more diffuse than if the figure had been perfectly true, yet the effect of those reflectors was very remarkable, and even comparable with that of concave mirrors; for bits of wood or cloth, held in the focus, were burnt through or set on fire, in a few seconds.

Specula cast of a hard bright composition, ground and polished, would no doubt answer every purpose still better; but to procure such, of any moderate size, would be to incur enormous expence. Besides, it will appear from the sequel, that the experiments for which the tin reflectors are peculiarly designed by no means require that elaborate accuracy of execution so indispensable in optical instruments, and depend for their success on no circumstance so much as the smoothness and high finish of the surface. As this is rather apt to tarnish, it is proper, from time to time, to refresh its lustre; which is easily done, by rubbing it with a cotton rag and a little chalk or whiting.

The

The parabolic figure is properly adapted for parallel rays, or such as proceed from a very remote object; yet it will answer, with tolerable exactness, in most other cases. The focus of parallel rays being once known, a very simple calculation will give that which corresponds to any particular distance.* Or it may be always found in this manner:—Place a small lighted taper at the radiant point, and having otherwise darkened the room, catch the reflected light on a slip of paper, and carry it forwards till the bright spot becomes most contracted and best defined. One reflector I had worked into an elliptical form, the distance between the two foci being four feet, and the nearer one only five inches from the apex. As it was the best suited to the distances which I found convenient, and was at the same time the most powerful in my possession, I employed it much more frequently than any of the rest.

When a reflector was used, it was supported from the table in an upright position by the help of a small wooden frame or stand, consisting of

* See Note II.

two narrow perpendicular pieces, extending somewhat above the centre of the reflector, but rather less distant asunder than its diameter, and morticed into a pretty broad horizontal piece with cross claws at the ends. On the inside, towards the top of each of those pillars, and near the middle of the flat piece, a slight groove was cut, through which the reflector was let down; and though it was held firm by the gentle spring of the wood, it could be easily removed at pleasure without deranging the stand.

To determine the action of heated bodies with any degree of precision, it is necessary to operate with large masses; for otherwise they cool so fast, as not to allow time for the regular and full production of their effects. Hot water seems to possess every requisite for that purpose; the facility of obtaining it, its great capacity for caloric, and the accuracy with which its temperature at every step of the process can be ascertained. To contain the water I preferred hollow cubes of block-tin, formed exactly, and planished, as the workmen term it, or hammered to a smooth and bright surface. These canisters had an orifice at the

the middle of the upper side, from half an inch to an inch in diameter, and the same in height, fitted to receive a cap through which was inserted a thermometer, whose bulb might reach nearly to the centre of the water. The cubes were of different sizes; of three, four, six, and ten inches. In two of them, namely, those of four and ten inches wide, the lid was not soldered, but could be adapted or removed at pleasure; and there being no occasion for an aperture, the stem of the thermometer was passed through a short pipe.

It was not without reason that I chose the cubical form. For, when any side was turned towards the reflector, every portion of the surface evidently presented the same inclination; it was easy likewise to ascertain how far different obliquities of position might affect the results; and there being four sides perfectly similar and equal, but whose surfaces could be variously altered, they afforded, without trouble, the means of multiplying the investigations in the same process. One side was constantly kept clean and bright, the opposite one was covered with writing paper pasted to it, or was painted over with a coat of lamp-black, mixed up with as little size as would make it take a body. The other sides, being allotted for miscellaneous service, were, according as the case required, coated indifferently with tin-foil, or coloured papers, or different pigments, or had the nature of their surface changed by mechanical or chemical agents.

The canisters were placed on light frames or stools with four feet, of such length that the middle of the canister stood at the same height from the table as the centre of the reflector. The frame was about half an inch narrower than the bottom of the canister, and had its edges rounded or chamfered. The square shape of the stools afforded this material advantage, that, by turning the whole round, the several sides of the canister could easily, in succession, be brought into the same identical position.

I had also, for the sake of variety, and the accommodation of some particular experiments, two similar cylindrial vessels made of tin, the one three inches in diameter and four inches high, and the other six inches in diameter and eight inches high; the latter having its top removable at pleasure. In the more advanced stages of my inquiry, I enlarged or altered the apparatus according as the circumstances suggested. Those additions or changes I shall notice in their proper place.

But the instrument most essential in this research, and to the superior delicacy of which I must in a great measure ascribe my success, was the *differential thermometer*. Nothing indeed could be simpler or more commodious than that which I used. Its general construction is the same as that of the hygrometer, of which a concise account has been already given to the public. Two glass tubes of unequal lengths, each terminating in a hollow ball and having their bores somewhat widened at the other ends, a small portion of sulphuric acid tinged with carmine being introduced into the ball of the longer tube, are joined together by the flame of a blow pipe, and afterwards bent into nearly the shape of the letter U, the one flexure being made just below the joining, where the small cavity facilitates the adjustment of the instrument, which, by a little dexterity, is performed by forcing with the heat of the hand a

few

few minute globles of air from the one ball into the other. The balls are blown as equal as the eye can judge, and from four-tenths to seven-tenths of an inch in diameter. The tubes are such as are drawn for mercurial thermometers, only with wider bores; that of the short one, and to which the scale is affixed, must have an exact calibre of a fiftieth or a sixtieth of an inch; the bore of the long tube need not be so regular, but should be visibly larger, as the coloured liquor will then move quicker under any impression. Each leg of the instrument is from three to six inches in height, and the balls are from two to four inches apart. The lower portion of the syphon is cemented at its middle to a slender wooden pillar inserted into a round or square bottom, and such that the balls stand on a level with the centre of the speculum. (See fig. 2, where the differential thermometer is represented at two-thirds of its natural size.) A moment's attention to the construction of this instrument will satisfy us that is affected only by the *difference* of heat in the corresponding balls, and is calculated to measure such difference with peculiar nicety. As long as both balls are of the

the same temperature, whatever this may be, the air contained in the one will have the same elasticity as that in the other, and consequently the intercluded coloured liquor, being thus pressed equally in opposite directions, must remain stationary. But if, for instance, the ball which holds a portion of the liquor be warmer than the other, the superior elasticity of the confined air will drive it forwards, and make it rise in the opposite branch above the zero, to an elevation proportional to the excess of elasticity or of heat. It is easy, after the mode practised in the case of the hygrometer, to fix the magnitude of the degrees for any particular instrument, and if it were expedient, other methods might be proposed which are applicable to the present instance.* The interval between freezing and boiling water being distinguished into an hundred equal parts, called *centigrade*, each of these subdivided decimally constitute the degrees which I employ, and which, following up the same system of nomenclature, would be termed *milligrade*. With the measures which I have stated, each dif-

* See Note III

ferential

ferential thermometer will contain from 100 to 150 degrees. I would observe, however, that such graduation is seldom positively required, and that, in most cases, it is less important to know the absolute quantities of heat than their relative proportions. I need scarcely add, that I had a variety of those differential thermometers, of different sizes, and of some diversity of forms, adapted for particular occasions.

CHAP.

## CHAPTER II.

IN a close room without a fire, place the tin reflector near the end of the table,* (See fig. 1.) and set the canister on its stand a few feet distant, and with its papered or blackened side directly fronting the reflector; and having, by means of a lighted taper or otherwise, found the place of the corresponding focus, move to that spot the ball of the differential thermometer containing the coloured liquor, which, to avoid circumlocution, I shall in future term the *focal* ball, and bring the plane of the instrument parallel to the face of the reflector. Things being in this state of preparation, fill the canister with boiling water, and adapt the cap with its thermometer. The coloured liquor of the differential thermometer will be perceived immediately to rise; in the space of

* The figure represents a screen in front of the canister, but which is introduced only in the experiments related in the next chapter.

two

two or three minutes it will have mounted near the top of the scale, and, having remained a short while stationary, it will afterwards slowly descend in proportion as the canister cools. I used commonly the six-inch canister, placed at the distance of three feet from the deep reflector; and, under such circumstances, the effect produced on the focal ball amounted at its highest range to about 80 degrees.* But after many trials, I found this effect, in every possible case, to be exactly proportioned to the heat of the canister, or the difference of its temperature from that of the room: an observation which, by introducing such simplicity, very much facilitated the prosecution of the experiments. The thermometer generally indicated 95 degrees centigrade, when I began to note the effect on the focal ball; and I continued at proper intervals to register the quantities, till the canister had cooled down to 50 or 60 degrees, so that a couple of hours perhaps elapsed before I had occasion to empty and refill it. From that register I calculated, by the rule of proportion, the quantities

* Equal to 14·$\frac{1}{2}$ degrees by Farenheit's scale.

which

which would correspond to 100 degrees of difference of temperature, or the whole interval between the freezing and the boiling points, which last numbers only I took the trouble to preserve.* And it would surely be preposterous to embarrass the attention of the reader with a multiplicity of figures and mere arithmetical computations: the facts which I have to state are not founded on the authority of single experiments, but are the mean results of numerous observations performed with the utmost care. Their coincidence was in general sufficiently striking, and if, in certain nice cases, any discrepancy occurred, I did never rest satisfied till, by frequent repetition, every doubt and uncertainty had disappeared.

But the experiments succeed equally with Cold as with Heat. If the canister be filled with ice, or with a frigorific mixture of snow and salt, the focal ball will be chilled, and the coloured liquor will consequently sink. The measure of the effect too, though in a contrary direction, is still rigorously proportional to the difference of temperature.

* See Note IV.

ture. Thus, if the liquor in the differential thermometer ascend forty-five divisions, while the temperature of the canister is 76 degrees, and that of the room 16 degrees; on filling the canister with broken pieces of ice, which will therefore continue at zero, the liquor will descend twelve divisions: but twelve is in the same proportion to sixteen that forty five is to sixty, or the difference between seventy-six and sixteen.

Those effects might be exhibited at greater distances, by employing two reflectors facing each other, and having their foci conjugate, the hot or cold body being placed in the one focus, and the sentient ball of the differential thermometer in the other. But this plan of experimenting, though not without beauty, is altogether unfit for any delicate inquiry: since, to obtain the peculiar advantage of that adaptation, the body suspended in the primary focus must be very small, and therefore its action will be only feeble and transient, incompatible with correct observation.

From what has been stated, it appears unquestionable that some hot or cold matter, according to the nature of the case, actually flows from the canister

canister towards the reflector, and from the reflector to the focal ball, where its impression accumulates till the complete effect is produced. Heat and cold, in every respect only relative, thus show the same measure of action, which must therefore be referred to the same identical cause. We have now to investigate what circumstances are capable of altering the energy of that emission. For the sake of distinctness, I shall, in this and some of the subsequent chapters, adopt one hundred to denote the extreme effect, or that of the blackened surface with the whole difference of temperature between boiling and freezing; and shall express the other quantities after the same proportion.

## EXPERIMENT I.

PAINT one side of the canister with lamp-black, coat another with writing-paper, and cover a third side with a pane of crown-glass of the same dimensions, fixing it down with pitch or hard cement. Thus prepared, dispose the apparatus for action, turn the black side of the canister to front the reflector, and fill it with boiling water. The liquor of the differential thermometer will rise to 100 degrees. Bring the papered side into the same position, and a similar effect, though rather smaller, will be produced, equal to 98 degrees. The vitreous surface will betray a sensible diminution, its action amounting to about 90 degrees.

Thus blacking, paper, and glass, constitute the same class of substances, whose effects, though somewhat different, are all very considerable.

## EXPERIMENT II.

Things being still in the same situation, direct the bright side of the canister to face the reflector, and the effect on the focal ball will be observed to suffer a very remarkable change, the coloured liquor quickly sinking to 12 degrees: but any side of the canister covered with tinfoil, and brought into the due position, will manifest precisely the same action. To produce the peculiar effect, it is only requisite to employ a clean metalline surface.

Thus, in its affection to heat, is tin radically distinguished from blacking, or paper, or even glass,

glass, since compared with them, it exhibits only about the eighth part of the energy. In both cases the heat of the presented surface must obviously be the same, and yet it is capable of producing at a distance effects so widely different. That such a difference of action could obtain between the metals and the soft or vitreous substances, our previous information would certainly not have led us to suspect. But, however paradoxical the fact may appear, it is not the less real and palpable. Nor is any nice apparatus required for detecting it; for, if I hold my hand about an inch from the blackened side of the canister, I feel a very sensible and agreeable warmth; but if I hold it at the same distance from the clear surface, I am scarcely conscious of any heat at all, till I bring my hand actually to touch the canister.

## EXPERIMENT III.

COVER the focal ball with a small bit of tinfoil, and make this to fit close all round, smoothing down the creases, but avoiding carefully to leave

 any

any scratches. Replace the differential thermometer, fill the canister again with boiling water, and present its blackened surface. The effect will now be reduced to about 20 degrees. Bring the clear side of the canister into the same position, and the effect will not exceed 2½ degrees.

Tin is, therefore, five times less susceptible of the impression of heat than glass, and thus the very remarkable distinction already noticed is conspicuous in the reception of heat, as well as in its propagation. Why there should be any difference in the proportion of those effects, will be the subject of future investigation.

## EXPERIMENT IV.

In the place of the tin reflector substitute a concave mirror, remove the cap of tinfoil from the focal ball, dispose the whole apparatus properly, and, having presented the canister's blackened side, re-fill it with boiling water. The coloured liquor will rise through a small but visible space. Rub off the silvering from the back of the mirror, and the effect will remain unaltered. Roughen the surface

surface of the back, by grinding it with sand or emery; the same effect will be still perceived; a decisive proof that the reflection of heat is produced entirely at the anterior surface of the mirror. Over that surface spread a body of china ink, which will form an even and glossy coat: replace the mirror, and the effect now becomes altogether insensible. Cover the face of the mirror with a sheet of tinfoil, by pasting and carefully adapting it to the curvature, and smooth away as much as possible the folds and rumples: a very great change will instantly be perceived in the degree of performance. The effect of this reflector will ten times exceed that produced by the naked mirror.

It hence appears that, independent of the polish and figure, the nature itself of the substance of which a surface consists, has a most predominant influence in determining the measure of the reflection of heat: indeed the former requisites are comparatively of much inferior consequence. Having lined a tin reflector with foil, I found that it still showed two-thirds, or even three-fourths, of the power which it had at first. But the very different effects of metallic and vitreous surfaces in the reflection of heat, cannot fail to strike the most indolent observer If I place a concave mirror at the remote end of a room, and opposite to a good fire, and hold my finger in the focus, I shall barely distinguish the heat collected there: whereas, if I set a tin reflector in the same spot, the heat accumulated at the focus will soon become intolerable, and greater even than what is felt within two or three inches of the fire itself.

It may be eligible to exhibit the several results of the foregoing experiments in a collective view. We shall thus perceive more clearly perhaps that, Heat, flowing from the same source, and acting at the same distance, can yet generate effects which are wonderfully different. Suppose that the bright side of the canister fronts a concave mirror, and that the focal ball, properly ranged, is coated with tinfoil; an effect very minute indeed, but under favourable circumstances, still perceptible, will be excited. Call this 1. Cover the face of the mirror with tinfoil, and the liquor of the differential thermometer will mount to 10. Remove the cap from the focal ball, and the action will be increased to 50. Now, present the blackened side of the

canister,

canister, and the extreme effect will be produced, equal to 400.

All those experiments succeed equally with Cold, which exhibits the same diversified effects, and after the same proportions, though necessarily more limited. Experiment II. cannot be performed satisfactorily, except in a dry state of the atmosphere; for otherwise the surface of the tin becomes quickly covered with dew, or crusted over with ice, either of which totally changes the measure of the effect. It must be confessed, however, that the experiments with Cold, though perfectly consonant, are much more troublesome in the execution, and require greater attention and stricter observation.

The facts related in this chapter will be deemed at least very curious; and viewed all together, they are calculated, I think, to affect us with surprize. Nay, they are repugnant to our first notions, and might experience contradiction, if they were not so easily verified. We might admit, perhaps, without much hesitation, that blacking, and even paper, are, by their constitution, more receptive of heat than the bright surface of metal;

but if this quality results from any particular affinity or superior attraction, how shall we conceive that those soft substances likewise discharge heat the most copiously? The power of absorbing heat, and the power of emitting it, seem always conjoined in the same degree; and this uniform conjunction clearly betrays a common origin, and discovers the evolution of a single fact, which assumes contrary but correlative aspects. In the reflecting of heat also, we readily perceive that the very different aptitudes exhibited by different surfaces are derived from the same principle. That portion of heat only is reflected which has not been previously absorbed. Thus, a coat of china ink affords no reflection perceptible, because it is most absorbent of heat. A concave mirror occasions a small degree of reflection, for its disposition to absorb, though very considerable, is manifestly inferior to that of blacking; and a surface of tin, as it retains very little of the heat, produces a most powerful reflection.

Whatever reasonings are employed concerning the operations of Heat, the same must, with equal propriety, apply to those of Cold. Do Heat and Cold

Cold constitute distinct elements, or are they only accidental and interchangeable qualities? Does Heat act upon remote bodies, by the flow of some peculiar species of matter? And does Cold exert its influence in a similar way, by the transmission of matter of an opposite kind? Or do not both of them produce their distant effects by the agency of the same individual fluid, susceptible, like all matter, of every possible degree of temperature?

## CHAPTER III.

PROVIDE a light frame of wood, wider than the diameter of the largest reflector, that is, about sixteen inches square, with feet to make it stand perpendicular. The purpose of it is merely to serve as a screen, having, as occasion requires, different thin substances attached to it.

### EXPERIMENT V.

OVER this frame extend a sheet of tinfoil, and having arranged the apparatus as before, the canister presenting its blackened surface, set the screen parallel to it, and advanced about two inches from it. (See fig. 1.) The effect upon the focal ball will now be completely intercepted; at least, if any impression be made at all, it is too minute to be discerned. And the same phenomenon occurs, whatever be the position of the screen between the canister and the reflector, if it be but sensibly detached from the former.

This

This striking experiment establishes incontrovertibly two points of essential importance towards grounding a Theory: 1. That there is an actual flow or impulsion of some corporeal substance from the surface of the canister, since the interposing of a sheet of tinfoil totally precludes the action of heat upon a distant body; and 2, That the matter of which this emission or undulatory motion consists is of a palpable nature, quite distinct from the subtlety and extreme tenuity usually ascribed to æther and other imaginary fluids; for the progress of the current or pulsation is absolutely stopped by a metallic plate which exceeds not the *five hundredth* part of an inch in thickness. Nay, I have found since, that this effect is produced by interposing gold-leaf, which is 600 times thinner, and indeed so amazingly thin, that notwithstanding the opaqueness of its substance, it will admit, in a very sensible degree, the passage of the rays of light.* The heated canister thus produces a direct impression merely in its immediate vicinity, and the effect is remotely communicated,

* See Note V.

or transferred only by the vehicle of a certain corporeal medium. Very different is the case with the action of gravity. By a primordial law of Nature, the mutual attraction of two bodies depends solely on their relative distances, and no barrier can in the least obstruct or impair its energy. A stone weighs exactly the same, whether above or below a table; nor is the gravitation of the earth to the sun at all affected by the intervention of the moon in a solar eclipse. The same phenomenon occurs in the action of magnetism. A loadstone will attract a bit of iron through a thick board, or a block of marble, with the same precise force which it would exert at that distance, if thofe seeming obstacles were removed. Electrical action manifests a like character.*

## EXPERIMENT VI.

Things being disposed as before; remove the tinfoil from the screen, and substitute in its place a pane of crown-glass. A very material change

* See Note VI.

will

will be now perceived. The liquor of the differential thermometer will rise to 20 degrees, estimating the entire and unobstructed effect of the blackened side at one hundred.

Thus, the intervention of glass does not, like that of tin, annihilate the effect upon the focal ball, but only reduces it to the fifth part of its former intensity. Tin is opaque, glass is diaphanous; has the fluid which is thrown from the surface of the canister any relation to light? To resolve this question, some farther considerations are necessary. Though light permeates glass and other diaphanous substances, it yet suffers in its passage a certain degree of diminution or absorption. That degree depends solely on the quality and the thickness of the transparent medium. Whether light passes in a condensed or a diffuse state, it must, in either case, sustain the same proportional loss; because each particle of which it is composed, travelling through the same range of matter, must incur the same risk of impediment. Here, then, is a simple criterion by which to decide whether the fluid, which is emitted from the heated surface, really penetrates through the substance

stance of the glass, and thence emerging, though with diminished quantity, continues its course; for it would experience absolutely the same measure of absorption, amounting to four-fifths of the whole, in whatever part of its transit, from the canister to the reflector, it encountered the screen.

## EXPERIMENT VII.

The apparatus still remaining in the same situation, carry the pane of glass successively forwards, keeping it constantly parallel and opposite to the blackened side of the canister. At each remove, the impression upon the focal ball will regularly diminish; insomuch that, when the screen has gained a position, one foot advanced from the canister, and consequently two feet from the reflector, it will not exceed the thirtieth part of the full effect.

The fluid thrown from the canister is not, therefore, like light, capable of permeating glass. But that it is of a nature perfectly distinct, will appear evident from the following experiment.

EXPE-

## EXPERIMENT VIII.

Remove the pane of glass, and in its stead attach to the frame a sheet of writing-paper; dispose the apparatus properly, and having placed the screen two inches before the blackened side of the canister, fill this with boiling water. The liquor of the differential thermometer will now rise to 23 degrees; an effect equal nearly to the fourth part of what is produced without the intervention of the screen.

Thus, with a screen of paper, which may be deemed an opaque substance, not only is the impression of heat conveyed, but even in a higher degree of intensity than when one of glass is used. The law too is the same, by which that impression diminishes in proportion as the screen recedes from the canister.

What then is this calorific and frigorific fluid after which we are enquiring? It is incapable of permeating solid substances. It cannot pass through tin, nor glass, nor paper. It is not light, it has no relation to æther, it bears no analogy to the fluids, real or imaginary, of magnetism and electricity. But why have recourse to invisible agents?

——————— Quod petis, hic est.

It is merely the ambient AIR.

But how shall we explain the diversified effects of different screens? By all of them, the current or pulsation of hot or cold air, in its progress towards the reflector, will be completely stopped: and, since the direct action of the canister is intercepted, the screen must operate by a secondary and derivative influence. From its position it acquires heat or cold, and, in its turn, displays the same energy as if it had formed the surface of a new canister of the corresponding temperature. It is no valid objection, that a substance so thin as paper, being incapable of containing much heat, is fitted only to produce a slight and fugitive effect. The screen is enabled to maintain its temperature, and consequently to continue its action, by the perpetual accessions of heat or cold which it receives from the canister.

It hence appears, that the quantity of effect produced upon the focal ball when screens are interposed

posed is determined by the combined operation of two kindred properties; their aptitude to receive heat, and their power to discharge it. Thus, with paper the effect is greater than with glass; because, as was formerly ascertained, the receptive and the dispersive qualities of the former are likewise greater. With tin no perceptible impression is made, for those qualities it has in a very inferior degree; and though of each taken singly the action might be discerned, the effect of their combined influence is too minute to be observed with certainty.

This explication is so clear and consistent, and seems to flow so naturally from the phænomena, as to carry with it irresistible conviction. To interpose a screen, amounts in fact to nothing more, than to substitute another canister with a certain correspondent but smaller charge of heat. The elevation of temperature which the screen acquires from its apposition, is the point on which its action entirely depends. Destroy or prevent this, and the effect immediately ceases. Nor is such an experiment impracticable, or even difficult to perform. A substance readily presents itself, which is so constituted as to be incapable, beyond a certain limit, of having its temperature raised.

## EXPERIMENT IX.

DISPOSE the apparatus as usual, detach the paper, and fasten to the frame a thin sheet of ice; the liquor of the differential thermometer, so far from rising, will now actually sink, and that in proportion to the warmth of the room. Take away the canister, and the effect will yet continue unaltered.

It is plain, therefore, that the ice here acts alone, and is not affected in any degree whatever by the proximity of the canister. The hot streams or pulses of air which play against it, expend their heat in melting its surface. The general temperature of the ice remains invariably the same; and even the sides, where the process of thawing goes on, continue at the point of zero, or very near it.*

* See Note VII.

EXPE-

## EXPERIMENT X.

SELECT two panes of crown-glass as flat and smooth as possible, and coat one side of each with tinfoil, by means of a little gum-water. Thus prepared, and the apparatus put in order, jóin those panes together with their tin surfaces in contact, and attach them to the frame of the screen; the focal ball will receive an impression equal to about 18 degrees. Invert the panes of glass, placing them with the tin coatings outmost: the liquor of the differential thermometer will now sink back again to the beginning of the scale.

Such is the *experimentum crucis.* It establishes beautifully and, I think, beyond the power of contradiction, the simple theory to which we have been led by a close train of induction. In both cases the obstacle presented, or the compound screen, is absolutely the same. If the effects in the focus of the reflector were produced by some subtle emanation capable of permeating solid substances, how could such a singular contrast obtain? It seems impossible to elude the force of this argument; but in a subject so curious, it may prove acceptable perhaps to relate a few more experiments, which tend at once to confirm and illustrate the same conclusion.

## EXPERIMENT XI.

CAUSE two sheets of tin about ten inches square to be hammered quite flat and smooth, and paint one side of each with a thin coat of lamp-black. Arrange the apparatus as usual, and having joined together the tin-plates with their clear surfaces touching, fix them to the vertical frame: the liquor of the differential thermometer will rise 23 degrees. Invert the position of the plates, so that the blackened sides come into contact: it will now sink down to zero. Remove either of the plates, and the liquor will again mount near 4 degrees.

It is truly pleasing to witness this varied spectacle, where the changes succeed each other as if performed by the fancied operation of magic. But those transitions, and even the measures of the

the diversified effects, are the necessary results of the principles already established.—Compare the case where both the external surfaces of the screen are metallic with that in which they are covered with pigment. On the one side it receives five times less heat, and this heat is propagated with eight times less energy from the other. By the joint influence of those circumstances, therefore, its effect is 40 times less; which corresponds to about half a degree, a quantity scarcely distinguishable. When the screen consists only of a single plate blackened on the one side, the diminished effect is a mean between the receptive and the projecting powers, or 6½ times smaller than where both surfaces are painted. This enfeebled impression is consequently equal to about 4 degrees.

## EXPERIMENT XII.

Affix those half-painted tin plates, with their bright side outwards, each to a separate frame. Dispose the apparatus as usual, and place the first screen with its blackened side fronting that of the

the canister at two inches distance. Bring the other screen into contact, the metallic surfaces joining: the focal ball, as it was formerly noted, will acquire a heat of 23 degrees. Now withdraw the second screen to a parallel position, two inches nearer the reflector, the effect on the differential thermometer will totally vanish.

When the two plates formed one body, they were evidently both of the same temperature; but after they were separated, the second screen derived its heat from the first by the medium of a double process, or the projecting and receptive powers of the opposite metallic surfaces: it had its excess of temperature above that of the room diminished therefore more than 40 times, and consequently its action on the focal ball proportionally impaired.

## EXPERIMENT XIII.

Procure some deal boards of different thicknesses, and planed on both sides, to act as screens: and matters being arranged as before, place them successively in the usual position. With a screen, one-

one-eighth of an inch thick, the effect will be 20 degrees; with another, three-eighths of an inch thick, it will be 15 degrees; and with one a whole inch thick, the effect will only be 9 degrees. Those quantities are not altered by painting the sides of the boards with lamp black. I need scarcely observe that in this experiment a very sensible time will elapse before the impressions are fully produced.

Here is indeed a successive diminution of effect, but it is extremely different from what would obtain if it were occasioned by the absorption which a fluid experiences in penetrating through successive lengths of passage. In that case a proportional loss would be sustained in the transit of each equal stratum, and the quantities of fluid transmitted would consequently form a descending geometrical progression. The effect corresponding to a screen one-eighth of an inch thick being 20 degrees, that corresponding to one of a quarter of an inch thick would be 4 degrees, and that corresponding to a thickness of three-eighths of an inch would be only four-fifths of a degree, instead of 15, which experiment gives. And pursuing the analogy, a board of an inch in thickness would not be capable of producing an impression of more than the four-hundredth part of a degree.

The variety of effect shown in the preceding experiment must be attributed wholly to the slowness with which heat or caloric pervades the substance of a thick board. In proportion to this thickness, the surface which receives the heat will acquire an elevation of temperature, and that which disperses it will suffer a similar depression. The inequality will increase till the heat is conducted through the internal mass as fast as it is dissipated at the posterior surface.

In screens, therefore, consisting of the same matter, the temperature acquired at the posterior surface will be directly as the excess of the temperature of the anterior surface, and inversely as the thickness of the substance itself. Thus, since the mean temperature in every case must be 24 degrees, when the screen of one-eighth of an inch thick was used, the temperature of the one side exceeding that of the room by 20 degrees, that of the other side must have had an excess of 28; those excesses in the screen of three-eighths of of an inch thick are 15 and 33; and those of the screen of one inch, are 9 and 39. The successive differences are 8, 18, 30; which being divided by the corresponding thicknesses, give 8, 6, 3¾; numbers proportional very nearly to 20, 15 and 9, which denote the power of the screens. In screens of the same thickness, but consisting of substances of different conducting powers for heat, the acquired temperature of the posterior surface will be directly as those conducting powers, and the difference between the temperatures of the two surfaces. Hence may be founded a convenient and elegant method of determining the conducting powers of solid bodies.*

I think it superfluous to pursue those illustrations farther. They corroborate in the clearest manner the principles before stated. Nor shall I have occasion again to repeat that they all succeed equally, and in the same exact proportions, with Cold as with Heat. Heat or Cold is propagated to a distance by the vehicle of the intervening air, and with very different degrees of energy, according to the nature of the surface at

* See Note VIII.

which the impulse originates, or upon which it terminates. These impulses must be transmitted in diverging lines, since the action of screens was observed to diminish in proportion as they receded from the source of heat. The velocity of transmission too must be prodigiously great, for though with a pair of bellows I blow strongly along the surface of the canister, I produce no sensible alteration in the focus of the reflector. Nor is it the exclusive prerogative of atmospheric air thus to convey heat; the other gases or elastic fluids, we shall find, possess the same property, though in various degrees, and with modifications peculiar to each. Are liquids capable of performing a similar office? This question, I presume, will be considered as resolved, if I discover what obtains in the case of water.

## EXPERIMENT XIV.

PLACE the apparatus within a large tub, and secure each separate part in its proper position. Fill the tub with cold water, so as to cover the whole, except a funnel soldered to the mouth of the

the canister. Things being thus disposed, pour boiling water into the canister; and whatever surface fronts the reflector, the differential thermometer will not be at all affected.

---

To every attentive reader it will now be apparent, that the theory of Radiant Heat, espoused of late years by chemical philosophers, is drawn from a very limited, vague, and imperfect knowledge of facts. Hence the total want of precision and the unavoidable obscurity and mystery in which it is involved. I will not stop to examine it. It is refuted by the whole train of the preceding deductions. But I am free to confess that the propagation of Heat is still a subject of immense difficulty. It strains the imagination to conceive by what singular process, a surface, in the act of cooling or heating, can dart its influence along with prodigious velocity on the wings of the ambient air. Are such motions compatible with the laws of fluids? If they really existed could they escape the cognizance of our senses?

But

—But why do I multiply objections? These will disappear in proportion as we advance. For the present, I shall content myself with taking a hasty glance of the mode of operation.

It is well known that bodies expand with heat, and that, in the case of the aëriform fluids those expansions are very considerable. And, since no absolute contact obtains in nature, Heat not only produces a dilatation among the particles themselves, but enlarges the limits that divide an elastic fluid from the solid by which it is bounded. As fast as the particles of air, therefore, come within the confines of the heated surface of the canister, they are forced to recede: nor is the equilibrium restored by the change of their position; receiving from their extreme proximity an impression at once sudden and complete, they acquire, in the space of receding, an impetus which thence displays itself in a continual propulsive or vibratory motion.

The effect of Cold is explained in a similar manner. The particles of air as they approach a cold surface, suffer a sudden contraction, and tend with force towards it; but this motion is opposed

posed by the surface, and reflected back with equal intensity.

That a *sudden* action is capable of exciting such effects in an elastic substance, may be illustrated by a familiar instance. If I compress a spring and then slowly withdraw my hand, it will passively follow, till it has regained its figure. But if I remove the pressure suddenly, the spring will recoil with violence, and perhaps continue to vibrate for some time. Similar consequences will ensue if I extend the spring, and leave it to collapse.

The diversified effects of different surfaces in the propagation of heat, may also be traced to the same source. A number of facts concur to show, that with glass, or paper, the physical contact of air is closer and more perfect than in the case of polished metals. The latter will therefore, it seems probable, exert a slower and more languid impression.

But how is this impression transmitted to a remote distance? Is it conveyed by propulsion and actual flight of the hot or cold particles of air? or is it communicated along the mass of

air

air by a successive but rapid transfer, accompanied with pulses or vibratory motions? We are not yet prepared for this very delicate investigation. We must first collect and examine a variety of new facts.

But whatever application of dynamics shall be found necessary to explain the recondite operations by which those singular phænomena are produced, our general positions must remain uncontrovertible. I am utterly at a loss indeed to imagine by what subterfuge it is possible to elude the force of the arguments which have been adduced. Are we again to be amused with the sportive freaks of some unknown intangible *aura?* Are the occult qualities of the schools to be revived and embodied in the shape of ætherial *media?* I have nothing more to urge. When prejudice retires behind an entrenchment of invisibles and possibilities, we must abandon the pursuit. It is vain to contend with phantoms.

When by the counsels of that illustrious martyr of science, Galileo, his ingenious disciple Torricelli performed the famous experiment of filling a sealed glass tube with mercury; the fact was so strikingly

strikingly beautiful, and so completely decisive of the weight of the atmosphere, that to barely mention the cause might seem sufficient to have opened with enthusiasm the eyes of the learned world. The adherents of the *fuga vacui* were indeed sadly perplexed, unable to reply, yet resolute to maintain their opinion. But their champion, Father Linus, defying the power of argument, very gravely asserted that the mercury was suspended from the top of the barometric tube by *invisible threads*. The good Father thus quieted the minds of the orthodox, and their generation slept in peace.

## CHAPTER IV.

ONE of the first steps towards discovering the nature of that aërial transmission or pulsation to the existence of which our inquiries have led, will be to ascertain in what manner its action is affected by the remoteness of its source. Of rectilineal emanations the power in general must evidently decrease as the squares of the distances from the radiant surface. But between the absolute quantity of effect and the degree of its intensity, there subsists a very material distinction. The same beam of light may aſſume a diffuse or a condensed form,—may cover a wide space with faint illumination, or act on a narrow spot with concentrated force. The quantity of light which the eye receives from a remote luminous body is diminished as the square of the distance; but, since the visual magnitude also decreases in the same proportion, its *brightness* or the intensity of its light

light must continue uniformly the same. In like manner, the number of rays that fall on the surface of a concave mirror is inversely proportional to the square of the distance of the radiant object: but, since the image subtends at the bottom of the mirror the same angle as does the radiant object itself, it must occupy a space in the same inverse duplicate ratio, and consequently, supposing the focus not to vary in position, it must have constantly the same degree of intensity. This is very nearly true when the object is remote, because, in that case, the correspondent focus does not sensibly alter its place. Nay, however paradoxical it may seem, the intensity of effect actually diminishes, though in a small degree, as the object is made to approach; for, by the laws of catoptrics, the focus then recedes somewhat from the mirror. And I may here state a theorem of considerable elegance: That, admitting the reflection of rays to be complete, their power of illumination in the focus of a concave mirror is exactly the same as what would obtain, if the surface of the mirror were converted into matter similar to that of the luminous source, and acted with direct energy.*

* See Note IX.

Thus, if a concave mirror be directed to the moon, the brightness at the focus will be the same, as if the mirror were removed and a circular surface of equal extent, but composed of lunar matter, were suppo ed to be substituted in its place. In like manner, if a portion of the sun's body could be transported hither, it would burn at the distance of the focus with the same intensity as a concave mirror of equal dimensions. Or, to employ an illustration more within our reach, a bit of paper, held to receive the image of the flame of a candle, will have the same degree of brightness, as if it were illuminated by a group of candles occupying the place of the mirror and covering with their united flames the same extent of surface. It is easy, therefore, to determine whether the äerial transmissions of heat or cold shoot along with unimpaired celerity and are subsequently reflected with the same precision that takes place in the mutual collision of elastic bodies or their impact against a hard surface. As long as the spot where the impressions are concentrated has sufficient extent to cover the ball of the differential thermometer, this must indicate the full intensity

intensity of action; and consequently, the measure thus obtained, when the canister is removed successively farther from the reflector, will continue undiminished, or rather will acquire a small increase. The only thing required, therefore, is, that the canifter, viewed from the bottom of the reflector, should in every case subtend as large an angle as the focal ball. The reflector which I used had a focal length of about 6 inches, and the balls of the differential thermometer were 4-10ths of an inch in diameter. Hence, if the distance of the canister did not 15 times exceed its breadth, the conditions were fulfilled. But to avoid every risk, I never placed a canister at a greater distance than 12 times its breadth.

## EXPERIMENT XV.

THE six-inch canister, presenting its blackened side, gave, at the distance of three feet, the standard effect of 100 degrees; but, moved back to the distance of six feet it produced an effect only of 57.

Thus, by placing the canister at double its former distance, the energy which it exerts is reduced to nearly one half. But had the effect been performed according to the laws of catoptrics, instead of 57 degrees, it would have amounted to 116; for 100 is to 116 as the square of 5.37 is to the square of 5.81, the focal lengths in inches corresponding to the distances of 6 and of 3 feet. And correcting the quantities in this way, I found in general, within the compass of my experiments, that the relative measure of effect was almost exactly as the reciprocal of the distance of the canister. This successive diminution cannot be imputed to any obstruction experienced in the passage through the air; for, in that case, it must have followed a very different progression. The effect at the distance of 3, 6, and 9 feet, instead of being denoted by the fractions $\frac{1}{3}$, $\frac{1}{6}$, and $\frac{1}{9}$, would have been expressed by the geometrical series $\frac{1}{3}$, $\frac{1}{9}$, $\frac{1}{27}$.

Such a striking deviation, therefore, from the properties of rectilineal emanations must originate somehow, either wholly or in part, from an imperfect reflection. Nor can it be ascribed to inaccuracy in the figure of the reflecting surface; for the

the focus being situate so near to the reflector, any defect of that sort must occasion a very trifling aberration. But should any suspicions be still entertained with respect to the influence of that source of error, they will be entirely removed by the following experiment.

## EXPERIMENT XVI.

INSTEAD of the tin reflector, I employed a very large concave mirror, of two feet in diameter, being the segment of a sphere of six feet radius. And, the heat of boiling water being hardly capable of making any visible impression, I preferred a charcoal fire, as it presented the most uniformly ignited surface, which was besides kept more regular by help of a constant and gentle stream of air from a pair of bellows. When the mirror stood at the distance of ten feet from the fire, the focal length being then four feet, the differential thermometer indicated 37 degrees. But after it was removed to the distance of thirty feet, the corresponding focal length being 38 inches, the effect produced was only 21 degrees.

To compare those effects exactly, it is necessary however to apply the correction due to the different focal lengths. As the square of 38 is to the square of 48,—or, in round numbers,—as 8 is to 5, so is 21 to 13. The action of the fire, at the distance of 30 feet from the mirror, if referred to the same focus as that at the distance of 10 feet, would therefore be 13 degrees; which is almost strictly the third part of 37, the real effect at the distance of 10 feet. And thus, as in the cafe of tin reflectors, the energy exerted is inversely as the distance from its source.

The principle which we have deduced from the known laws of catoptrics, was beautifully confirmed by the *Photometer*. This instrument, contrived to measure the intensity of light, will be fully described hereafter. I will just observe by the way, that it is merely the differential thermometer under a peculiar compact form, with one ball black, and the whole included within a glass case. It is therefore affected by light only, which being admitted through the cafe, acts from abforption on the black ball. When the concave mirror was 10 feet distant from the charcoal fire, the photometer

photometer marked 50 degrees, while the simple differential thermometer indicated 37 : but after it was removed to the distance of 30 feet, the photometer roſe to 78 degrees, and the differential thermometer at the ſame time declined to 21. It has been remarked already that the intensities corresponding to thoſe different foci ought to be in the ratio of 5 to 8. This would give 80, instead of 78, for the effect on the photometer at the distance of 30 feet. But the agreement is as accurate as could be reasonably expected in an experiment of such a nature. A striking contrast is exhibited between the reflection of light and that of heat.

But to return to the observations made by the tin reflector, which possesses so many advantages.

## EXPERIMENT XVII.

THE same standard being assumed as before, a canister of 3 inches square, with its blackened side fronting the reflector and 3 feet distant from it, produced an effect of 50 degrees; one 4 inches square, at the distance of 4 feet, gave 54 degrees;

one 6 inches square, at the distance of 6 feet, gave 57 degrees; and another canister 10 inches square, and 10 feet distance, gave 59 degrees.

These quantities are nearly equal; but if the proper correction be made for the different focal lengths, their identity will become apparent. The focal lengths corresponding to the distances of 3, 4, 6, and 10 feet, are respectively 5.81, 5.58, 5.37, and 5.22 inches. Their squares are 33.8, 31.1, 28.9, and 27.3; and reducing the numbers obtained by the experiment in the same proportion, the true effect at 4 feet distance will be 51, that at 6 feet, 49, and that at 10 feet, 48. The small differences that occur among these results are probably owing to the imperfection of the experiment, which was only repeated once.

It is plain, from their arrangement, that the canisters all subtended the same angle, and consequently that their energies were conveyed in the same lines, and, striking the same points of the reflector, were sent back by the same identical routes. This series of experiments must therefore have all been alike affected by the process of reflection; and whatever deficiency might arise from

from that source, it must have taken place and to the same degree in each. And as the numbers finally obtained were equal, or very nearly equal, we may conclude that the remote impressions of heat or cold are conveyed without any sensible diminution from the various length of their paſſage through the air. In this respect at least, the transmission of heat will admit of comparison with that of light, which suffers by absorption the loss of only one-fifth part in its perpendicular descent through the whole atmosphere.

It seems then ascertained that, within moderate limits, the action on the focal ball is proportional to the angle which the canister subtends. the impression consequently becomes more intense as the focal image enlarges. But if it is enlarged in the proportion of the visual angle, the whole quantity of effect would evidently be likewise attenuated in the same ratio, and of course the intensity would continue unaltered; that enlargement must therefore follow a slower progression than the angle which the canister subtends. We are hence led to this simple conclusion, that the reflection of heat is liable to be affected

affected with a certain constant measure of aberration or dispersion. Thus, besides the principal reflected ray, there are others extending perhaps 10 or 15 degrees on either side of it, but growing rapidly feebler as they diverge. But this declining expansive aberration may be considered as equivalent to an uniform aberration within the limit of 5 degrees. The focal image, augmented by this additional ring or rim of 5 degrees in breadth, forms therefore a space over which the action is in ordinary cases similarly diffused. Its intensity is hence inversely as that space compared with the visual magnitude of the canister. To illustrate this, let me observe that a canister of 3 inches, at 3 feet distance, subtends an angle of about 5 degrees. Call the breadth of its optical focus 1, that of another corresponding to a canister of 6 inches at the same distance will be 2; the breadths of the enlarged foci will therefore be as 3 to 4, and their spaces as 9 to 16. But the respective quantities of incident heat from the canisters are as 9 to 36; consequently the power of the 6 inch canister is increased in the ratio of 36 to 16, or is 2¼ times as great as that of of the 3 inch one; which differs little from experiment. In the same manner, comparing the action of a 3 with a 9 inch canister at the former distance: the breadths of the enlarged or igneous foci are as 3 to 5, and their spaces as 9 to 25; but the measures of the incident heat are as 9 to 81; whence the intensity of effect with the 9 inch canister is $\frac{81}{25}$ or 3 $\frac{6}{25}$ times greater than what is produced by the 3 inch one. Again, comparing the same standard with a 12 inch canister, the spaces of the igneous foci are as 9 to 36, and the sides of the canisters are as 9 to 144; wherefore the power of the 12 inch one is augmented 4 times, agreeable to observation. But this mode of computation cannot safely be pushed much farther; for after the surface of the canister becomes very broad, the parts towards its extremities cease to mingle their action with that of the central portion. On the other hand, if the heated surface be either very much contracted or removed to a very great distance, the optical focus will vanish in comparison of the igneous focus, and the effect will be simply as the visual magnitude. Hence the impression caused even by the 10 inch canister

nister will be barely sensible at the distance of 100 feet.

This subject will admit of being illustrated somewhat differently. Suppose, for example, that we were to compare the effects of a 3 and a 6 inch canister at the same distance from the reflector. The square of 6 inches may be distinguished into a central square of 3 inches, with a surrounding space of 1½ inch in breadth. This space will act partly by direct, and partly by oblique impression on either side. And it seems probable that the energy is thus equally divided: one third part of it is therefore added to the central square; but the third of 27, the quantity of outer space, is 9, which joined to 9, exactly doubles the effect.

The existence of this remarkable kind of aberration which takes place in the reflection of heat is further established by the following experiment.

EXPERI-

## EXPERIMENT XVIII.

The six inch canister was placed with its blackened side fronting the reflector, at the usual distance of three feet, and the differential thermometer adapted to its place. A small taper of the same height as the middle of the canister was gradually moved along its surface, and the progress of the reflected image at the same time marked. The taper was scarcely drawn aside two inches from the axis of the reflector, when the luminous spot had completely left the focal ball. Such then is the extreme limit of the optical focus. But the action of heat was much more extensive. Removed by successive ſtations along the table at right angles to the axis of the reflector, the canister still continued to exert its effect, though with declining energy, on the differential thermometer; nor did the impression cease to be distinctly perceptible till the canister was drawn aside 7 or 8 inches. At the first inch of displacement, the action was unaltered; in the next it was somewhat abridged; but afterwards it rapidly

pidly diminished. The numbers were nearly as follow.

| Position | Power |
|---|---|
| In the axis - - | 100 |
| 1 inch aside - - | 100 |
| 2 do. - - - | 83 |
| 3 do. - - - | 58 |
| 4 do. - - - | 33 |
| 5 do. - - - | 16 |
| 6 do. - - - | 4 |
| 7 do. - - - | 1 |

The law of this series is not very obvious. The terms are proportional to the fraction $\frac{5}{6}$ raised successively to 1st, 3d, 6th, and 10th powers, these exponents being the triangular numbers. From this series all the phænomena noticed above may without much difficulty be deduced; but I consider it unnecessary to dwell longer on the subject.

A curious consequence derivable from the aberration which affects the reflection of heat, is that the effect does not attain its maximum at the true focus, but somewhat nearer the reflector. To

To explain this perspicuously it will be expedient to borrow the assistance of a diagram. Let AB (fig. 6.) be a reflector, AD its axis, and F the true or optical focus. Besides the principal reflected ray BF, there are other indirect ones between the limits of the equal angles CBF, and DBF. But from a well-known proposition in elementary geometry, CF is to FD as CB is to BD, and consequently FD is greater than CF Therefore the heat received on FD is more diffuse and attenuated than that which falls on CF. Hence the ball of the differential thermometer, occupying a small space all round, and participating both of the direct and dispersed heat, will be less affected at F, than at some point *f*, nearer the dense portion of heat.

Experiment confirms that result. Advancing the differential thermometer half an inch nearer the reflector than the true focus, the effect was augmented more than one third; and bringing it still closer by half an inch, though the effect began to decline, it was still one fourth greater than at first. On the contrary, when the differential thermometer was moved backwards, the impression

impression diminished very fast, and at one inch beyond the focus, it did not amount to the seventh part. One example will suffice.

## EXPERIMENT XIX.

The six inch canister was placed at the distance of six feet from the reflector, and the differential thermometer was brought exactly to the optical focus. The effect was 58 degrees. Half an inch nearer the reflector, this rose to 80 degrees, and one inch nearer it was still 70 degrees. But drawing it successively from the reflector, at half an inch beyond the focus the action was only 25 degrees, and half an inch farther it sunk to 8 degrees.

## CHAPTER V.

IT has been already shown, that the original impressions of heat and cold are not propagated merely in lines perpendicular to the side of the canister. But this may be rendered obvious by giving the blackened surface a small degree of obliquity in regard to the axis of the reflector, for the effect is thereby not visibly altered. Those impressions are, therefore, conveyed in diverging lines. Are they likewise diffused equally in all directions? Such is supposed to be the case with the rays of light in flowing from a luminous body. But though analogy might induce us to extend the principle, it cannot be safely admitted without investigation. Fortunately, a very simple method presents itself for resolving this question. In all uniform radiations, the force of the rays from each single point is, at equal distances, the same, in whatever direction they issue; and consequently their aggregate effect must depend entirely upon the number of those points, without being in any respect modified by the relative position or the inclination of the radiant surface.

## EXPERIMENT XX.

SET the canister at a distance from the reflector not less than ten times its breadth, and dispose the apparatus as usual. In this position, the action of the whole of the blackened surface will be concentrated upon the focal ball. Turn the side of the canister successively more and more oblique, keeping its centre however always in the same place. The corresponding effect will continually diminish; at first gradually, and afterwards with accelerating activity.

The impressions of heat or cold are, therefore propagated through the air, with unequal degrees of diffusion. Their force is evidently greatest in the line perpendicular to the surface, and regularly decreases as the direction becomes more oblique. Between that force and the angle of obliquity

obliquity some relation must subsist; which is the next object of inquiry.

## EXPERIMENT XXI.

PROVIDE a tin screen, composed of two sliding parts that shut together in a vertical line, but leave, when opened, an aperture or slit, of any required breadth. Arrange the apparatus as formerly, and plant the screen a little before the canister and parallel to its blackened side. Open the screen, by drawing out both slides equally; and note the effect produced upon the differential thermometer. Now, turn the side of the canister about its centre, till it is just sufficient to fill up the void space behind, or such that no straight line can pass, by the edge of the canister and through the aperture, to the surface of the reflector. As the aperture of the screen is thus successively contracted, the canister will acquire an inclination always more oblique. In every case, the impression made upon the focal ball will depend on the quantity of aperture, and will be nearly the same, whether the canister stands parallel or inclined to the screen. When the obliquity becomes indeed very considerable, a small diminution of effect, seldom amounting to the tenth or twentieth part, begins to be perceived.

In this statement, I have purposely omitted to mention the numerical results, because the reduction of them would lead to a tedious and obscure discussion. If the reflector were removed to an indefinite distance, the lines proceeding towards it, from the blackened side of the canister, might be deemed parallel; and consequently the portion of that surface whose action is exerted, would be accurately defined, by two planes from the edges of the aperture at right angles to the screen. But those boundaries are very sensibly enlarged, in consequence of the angle which the reflector subtends. When the face of the canister is parallel to the screen, an appendage, or sort of penumbra, is annexed on either side to the proper limits, which is marked out by lines from the extreme edges of the reflector, and which produces a certain partial effect. When the canister has an oblique position, that penumbra, though of double breadth, occurs only on the one side; but as it is rather more

more remote, and makes a much acuter angle with a line drawn to the middle of the reflector, its auxiliary impression is palpably diminished. From the examination of some particular facts, I am disposed to balance that deficiency against what was above remarked. We may therefore conclude in general, that the remote action of a heated surface is equivalent to that of its orthographic projection, or can be estimated by the visual magnitude of its source. Hence a canister of a prismatic form, and having its acute angle turned towards the reflector, will produce the same impression upon the focal ball, as if, with an inverted position, it presented its base. This experiment I have not tried, but I have made another which in some respects possesses superior advantages, since it exhibits the combined effect of every possible inclination of surface.

## EXPERIMENT XXII.

AT the distance of five feet in front of the reflector I placed a cylindrical canister 6 inches in diameter and the same in height, having its anterior surface painted with lamp-black. The effect on the differential thermometer was noted, and this canister being removed, another of the same dimensions, but of a cubical shape, was substituted in the same position, or rather about an inch farther back. The same effect was still produced.

The several degrees of force corresponding to the various inclinations of the heated surface may be also investigated in a manner somewhat different, and which supersedes the application of screens. This it will be proper to relate.

## EXPERIMENT XXIII.

Having painted one side of a square canister and refreshed the metallic lustre of the rest, place it at a proper distance from the reflector and adapt the apparatus. Turn the blackened surface successively round on its centre, and mark at each interval the impression made on the focal ball. The approximate effects corresponding to different degrees of obliquity will be thus obtained. But they will always exceed the true values; for the adjacent metallic side of the canister will, from aberration,

aberration, mingle its influence with the principal action. To detect the error arising from that cause, at each successive station turn the canister one-quarter round, so that the place of the blackened side may be exactly filled by the bright one next it. Both cases evidently are alike affected by a lateral metallic surface, and the latter results being severally subtracted from the former, must exhibit the excess of action by the blackened side and above that of its succeeding bright one, with the same inclination to the axis of the reflector. Augment the last numbers, therefore, by one-seventh part, and we ſhall obtain the absolute measure of effect corresponding to various obliquities.

All the preceding investigations concur to establish this simple proposition—*That the action of a heated surface is proportional to the sine of its inclination.* To illustrate this: Let AB (fig. 7.) denote the position of the heated surface, draw the perpendicular AC, and describe the quadrant CDB. The power to transmit heat by the vehicle of the atmosphere to D in the direction AD is as DE, the sine of the angle of inclination DAB. Or, having described the semicircle CFA,—that power will

 be

be expressed, and perhaps more elegantly, by the intercepted portion AF, which is easily proved to be equal to DE, for the triangle ADE is equal in every respect to CAF.—But the comparative effects belonging to various degrees of obliquity, may be conveniently exhibited to the senses by a single diagram. Suppose AB, (fig. 8.) as before, represents the position of the side of the canister; describe the semicircle ADB, draw the perpendicular CD, on which describe an inner circle. Let the radii CH and CG, CE and CF, on either side of the perpendicular, denote several directions in which the impressions are conveyed: then will the intercepted parts C*h* and C*g*, C*e* and C*f* respectively denote the intensity of action propagated along those lines.

A number of curious corollaries flow immediately from the proposition now stated; but I shall not stop to notice them. I will only observe that the action in front of the canister is exactly double what would have been exerted, if the law of uniform radiation had obtained. For on this supposition the energies would be equally diffused over the surface of an hemisphere instead of its ortho-

orthographic projection; but the surface of an hemisphere, it is well known, is double that of its base.

The same principle receives additional confirmation, from the experiments made by interposing a paper screen at different distances before the canister. The nature of these has been already pointed out. It was shewn that the screen is not merely passive, but performs an active part; and that it operates as a secondary canister, with a force proportional to the temperature which it acquires from its apposition to the primary source of heat. To determine the heat communicated from the canister to the screen is a problem in the higher geometry, which, with the preceding data, is capable, under certain admissible limitations, of a simple and neat resolution.* The heat so conveyed is in every case as the square of the sine of the angle subtended at the centre of the screen by the semidiameter or half the breadth of the canister. With the same canister, therefore, it may be estimated by the inverse ratio of the distance of the middle of the screen from

* See Note X.

the edge of the canister. Thus, the several degrees of heat which the blackened side of a canister ten inches square can induce upon a screen at the successive distances of 1, 2, 3, 4, 5, 6, 7, 8, 9, 10, and 20 inches are respectively denoted by the fractions, $\frac{1}{26}$, $\frac{1}{29}$, $\frac{1}{34}$, $\frac{1}{41}$, $\frac{1}{50}$, $\frac{1}{61}$, $\frac{1}{74}$, $\frac{1}{89}$, $\frac{1}{106}$, $\frac{1}{125}$, and $\frac{1}{425}$. The progression declines at first very slowly, but, in the remote terms, it constantly approaches to the duplicate ratio of the distance.

## EXPERIMENT XXIV.

HAVING placed the ten-inch canister at the distance of $7\frac{1}{2}$ feet from the canister and duly arranged the apparatus, I planted a paper screen of 16 inches square directly in front of the blackened surface, and successively at the interval of 1, 2, 5, 10, and 20 inches, allowing the space of ten or fifteen minutes at each move. The corresponding effects on the focal ball, reckoning the full impression without the intervention of the screen at 100 degrees, were observed to be 32, 28, 18, 8, and 3.

According

According to theory, those numbers ought to be proportional to the fractions, $\frac{1}{26}$, $\frac{1}{29}$, $\frac{1}{50}$, $\frac{1}{125}$, and $\frac{1}{425}$. This gives the series 32, $28\frac{3}{4}$, $16\frac{1}{2}$, $6\frac{1}{2}$, and 2; the agreement of which with the observed quantities is abundantly satisfactory. But the coincidence will be still more striking, if the proper correction be introduced. In consequence of the approximation of the screen to the reflector, its action on the differential thermometer will, as formerly remarked, be proportionally augmented. Hence in the present instance, when the screen stands five inches before the canister, there is an addition due of $\frac{1}{17}$; at ten inches, $\frac{1}{8}$; and at twenty inches, $\frac{2}{7}$. The corrected numbers are 32, $29\frac{1}{4}$, $17\frac{1}{2}$, $7\frac{1}{4}$, and $2\frac{1}{2}$.

CHAP-

## CHAPTER VI.

HAVING examined at some length the general properties of those Aërial Transmissions or Pulsations of Heat or Cold, it remains to determine in what degree they appear to be affected by the species and quality of the propellent surface. The surface from which their impulsion originates may be considered under five principal points of view: 1. The nature of the substance of which it consists; 2. Its condition with respect to polish or lustre; 3. Its thickness; 4. Its disposition to hardness or softness; and 5. Its colour. Other circumstances may possibly modify the results; but the subject already embraces a sufficient extent. Nor will I pretend to exhaust or even to discuss it, with the same minute attention that has appeared in some other parts of our inquiry. I shall content myself with stating a few leading facts, which afford materials for theory.

theory. And I may perhaps indulge a hope, that the curiosity of some of my readers will be excited to prosecute those experiments in a fuller manner. I proceed to a concise enumeration.

1. The nature of the propellent surface is only the aggregate of its qualities, physical and chemical. Could we distinguish the separate influence of each, it would be easy to determine the result of their conjoined operations. But this developement perhaps exceeds the human faculties, and we must rest satisfied with a more humble and vague survey. It has repeatedly been remarked that the metals, compared with glass or paper, possess, in a very inferior degree, the power of transmitting, by the vehicle of the ambient air, the impressions of heat or cold. A considerable diversity in that respect, though confined within much narrower limits, will be found to obtain likewise among the metals themselves. It requires some nicety, however, to ascertain the scale of effects, as the impressions produced by metals are generally so very small. I thought it unnecessary to have canisters formed of different materials, nor was it easy to procure thin metallic plates

plates of the proper dimensions to apply successively to the heated surface. The want of better apparatus I endeavoured to supply by expedients.

## EXPERIMENT XXV.

I APPLIED the broad plate of a saw to the front of the canister, and securing it in that position, I observed the impression made on the focal ball. I then painted the outer surface with lamp-black, and noticed the effect now produced. Reckoning this equal to 100 degrees, the former amounted to 15.

The effect of a bright surface of tin in the same situation is only 12. We may, therefore, infer, that the propellent power of iron or steel is one-fourth more than that of tin.

## EXPERIMENT XXVI.

THE blackened side of the canister producing an effect of 100 degrees; another side where the tin was tarnished slightly with quicksilver and of an uniform matt white, was brought to face the reflector.

reflector. The effect was now reduced to 14 degrees. But another side, profusely moistened with quicksilver, which formed a resplendent and almost fluid coat, again increased it to about 20 degrees.

It is probable, therefore, that if quicksilver could be commodiously tried alone, it would cause double the impression of tin.

As the surface of tin becomes tarnished by exposure to the weather, this incipient oxydation increases its propellent power. And, in general, the aptitude of an oxyd to thow off heat is proportioned to the interval by which it recedes from the metallic nature and approaches to the condition of earths or vegetable substances. The standard effect of a coat of lamp-black being 100 degrees; that of a clean but rough surface of lead was 19; that of another sheet of the same metal, which, by long exposure, had acquired a dark grey crust, was 45; that of black lead or plumbago was 75; and that of red lead or minium was 80. The side of the canister being covered with a thick crust of dry size or isinglass, the effect was also 80. Hence it is of consequence to pay attention

tion to the quantity of size employed to give a body to the lamp-black, which forms the standard coating to the canister. As little should be used as possible, else the propellent power of the pigment will be considerably diminished. Thus, when the lamp-black was mixed up with a large proportion of size, the effect on the differential thermometer was only 90 degrees. For the same reason, China ink, which owes its fine glossiness to a large admixture of vegetable glue or gum, was found to cause, instead of 100, an effect only of 88 degrees.

The propellent powers of sealing-wax and rosin are almost equal and nearly at the limit of the standard: that of the former is 95, and that of the latter 96. Writing paper produces an effect equal to 98 degrees. Of the relative property of water I can only judge indirectly; but it is certainly very great, and perhaps exceeds that of lamp-black itself. Ice, I was enabled to try, by filling the canister with a freezing mixture of snow and salt, and moistening the outside repeatedly with water by help of a brush. A solid crust was thus soon formed. Its effect amounted to about 85 degrees.

2. The quality of a surface with respect to smoothness and polish, appears in certain cases to have a remarkable and very singular influence on the degree of its propellent power. The action of glass, or paper, or blacking, is not perceptibly modified by destroying their superficial gloss. But towards the other extremity of the chain, such a process occasions a most striking alteration. By filling a metallic surface with numerous parallel *striæ* or slender furrows, its effect in discharging heat may be more than doubled.

## EXPERIMENT XXVII.

The power of the blackened side of a canister being denoted by 100, that of a clear side as before, was 12. Another side, which had been slightly tarnished, was scraped to a bright but irregular surface: The effect was now 16. Another side was ploughed in one direction by means of a small toothed plane-iron, used in veneering, the intervals between the teeth being about the $\frac{1}{30}$ or $\frac{1}{50}$ part of an inch: The effect was farther increased to 19. The first smooth side was now scraped downwards with the edge of a fine file: Its power was 23. But the filing being repeated, and more thoroughly covering the surface; the effect rose to 26.

These few trials are sufficient to establish the fact, which is certainly very curious and unexpected. It is manifest that the propellent power increases in proportion as the scratches or *striæ* become multiplied on the surface. There is no doubt some limit where the effect reaches its maximum, and which might be discovered, by having plates of metal manufactured with a variety of delicate flutings. But the general experiment may be performed with greater facility in another way. Applying tin-foil, it will adapt itself exactly to the side of the canister: Its effect is 12, the same as that of polished tin. Then rubbing it in one direction with a bit of sand-paper, its surface will be covered with parallel strokes or scratches. The effect is thus augmented to 22. Fine sand-paper will be found to answer the best; and the furrows which it makes probably do not exceed the $\frac{1}{500}$ part of an inch in depth.

It was shown at an early stage of our inquiry that

that the power of a substance to discharge heat is collateral and equivalent to that with which it receives heat. Hence we derive a simple and convincing method of displaying the singular property of a striated metallic surface.

## EXPERIMENT XXVIII.

COVER each ball of a differential thermometer neatly with a coat of tinfoil, and rub that one below which the scale is affixed gently with sand-paper; or it may be rubbed before it is applied to the glass. Placing the instrument now in the sun, the liquor will visibly rise, perhaps 5 or 10 degrees: And the reason is obvious; for the light is reflected more copiously from the bright surface of the tinfoil which had been rubbed, and of course it is absorbed in a smaller proportion; consequently the propellent elasticity of the heat excited in the other ball causes an elevation of the coloured liquor. Set this differential thermometer now directly opposite to the fire, and about two or three feet distant from it. In this situation, a very remarkable depression will quickly take place, equal perhaps to 30 or 40 degrees. Interchange the metallic coating, and the same effects will be produced but in a reverse order.

This beautiful experiment likewise indicates clearly the distinction between the solar rays and culinary heat. Heat is not transmitted immediately from the sun, but is capable of being excited by light, and this in proportion to the degree of absorption which takes place. The light from the fire has some tendency to counteract or diminish, in a certain measure, the peculiar effect of the heat emitted from the same source.

Some will perhaps be disposed to refer the superior efficacy of a striated plate to its augmented quantity of surface. But it has been proved that the propellent, and, by analogy, the absorbent, power of a surface is proportional to the sine of the angle of its position. The action, therefore, which it exerts muſt depend on its visual magnitude or its orthographic projection. If those furrows enlarge the measure of surface, this effect is exactly counterbalanced by the obliquity of the contours which they present to the reflector. The action of those *ſtriæ* is besides proportionally greater than the augmentation

mentation of superficial volume. Nay, if the scratches be furrowed across, the power is again diminished, and reduced nearly to that of a plane surface.

3. The term *surface* is used, throughout this discourse, in its physical, and not its mathematical, acceptation. I employ it to signify a stratum of matter of a certain finite depth, yet of such extreme tenuity as almost to escape the cognizance of the senses. Nor is it one of the least interesting parts of our inquiry, to discover the degree of influence, in modifying the principal effect, which the various thickness of the superficial plate is capable of producing. But the experiment requires *laminæ* of such wonderful subtility that they can be seldom exhibited under a detached and independent form. We are left therefore to the exclusive choice of two methods: 1. To cover glass or paper with an extremely attenuated sheet of metal; or 2. to coat planished tin with a fine pellicle of animal or vegetable substance. The former is beset with difficulties, and is besides hardly practicable to the desired extent. The latter has fortunately, from its facility and delicate accuracy,

every advantage to recommend it. If a solution of glue or isinglass be spred upon a smooth level surface, it will congeal and afterwards dry into a fine pellucid coat of uniform consistence, resembling talc in its appearance. By cutting it into parallel slips and joining a number of them together, we may easily determine its exact thickness; and inverting the process, we may, with equal facility, compute the number and size of those slips, which, when redissolved and spred over a given extent of surface, are necessary to form a pellicle of the required tenuity. My first object of inquiry was not to ascertain the absolute but the proportional quantities. As I proceeded I sought to estimate, with scrupulous precision, the thickness of the pellicle. Having prepared a broad plate or cake of isinglass, of a horny consistence and about one-sixtieth of an inch thick, I cut it into small squares or oblongs to suit the calculated dimensions. Those subdivisons might also be performed by weighing; but nothing is so tedious or discouraging as the management of a fine balance.

EXPERIMENT

## EXPERIMENT XXIX.

On a bright polished side of the canister I spred a small portion of liquefied jelly, and quadruple that quantity afterwards on another similar side. Both coats dried into remarkably thin films, the first being iridescent, or unfolding those delicate mutable colours exhibited in feathers and soap-bubbles. Disposing the apparatusas usual, the effect of the blackened side of the canister being reckoned 100 degrees; that of the thinest film was only 38, and that of the other, which had four times its thickness, 54 degrees.

We have already seen that a thick layer of isinglass produces an effect equal to about 80 degrees; when attenuated, however, it suffers a visible diminution of power. The separate impression of the substratum of polished tin in the experiment was 12 degrees. Hence the influence of the one film may be estimated at 26, and that of the other at 42 degrees; which thus increases with their thickness, though not in the same ratio. After the superficial plate of isinglass has acquired a thickness of about the thousandth part of an inch, its effect is not perceptibly augmented by any subsequent additions of matter.

## EXPERIMENT XXX.

I HAD a cylindrical vessel, six inches in diameter and eight inches high, made of ordinary tin not planished, and with a surface tolerably bright but uneven. One side of this I painted with lamp-black; and arranged the apparatus for action. Reckoning as before, the effect of the blackened side to be 100 degrees; the vessel being turned half way round, to make the bright side front the reflector, its action was 25 degrees. Rubbing this metallic surface with a feather dipt in olive-oil, and allowing some time for the oil to flow off, the effect was now 51 degrees. But applying more oil, and noticing as early as possible its impression on the differential thermometer, the effect was found to be augmented to 55, or even 59 degrees. The heat of the canister, during the operation, was nearly 80 degrees of the centigrade scale

This

This experiment is of such an obvious nature as scarcely to require any comment. The film of oil was not of extreme tenuity, since it was not iridescent. We may remark, by the way, the very considerable propellent power of common tin, amounting to 25 degrees. That of planished block tin we have seen, is only 12, or at most 13 degrees. Such is the influence of a rough irregular surface!

I might here mention some other facts which concur to prove that the propellent quality of metal is affected by the thickness of its superstratum. Thus, filling the canister with a freezing mixture of snow and salt, the surface of the tin becomes covered over with hoar-frost, and as the fine icicles accumulate, the action on the focal ball is likewise increased. From 20 degrees, at which the liquor stood when first observed, it successively mounted to 90. Moistening the hoary crust with brine or a strong solution of common salt, which stopped the congelation and left a thin liquid film, the effect was reduced to 43 degrees, and continued stationary.

4. The disposition of a substance to discharge

or

or absorb heat appears to bear some relation to its degree of *softness*. Thus, lead has more efficacy than tin, and paper more than glass. But the qualities of hardness or softness are not of a very definite nature. Between solid and fluid bodies there indeed exists no absolute limit; and a continued chain might be traced, though all the intermediate degrees, from extreme hardness to the softness which constitutes a perfect liquid. I may on a future occasion enlarge on this topic, and produce several new and curious facts connected with it. At present I confine myself to a very general statement. The principle is illustrated by the gradual melting of wax, and the softening of oil, by the application of heat. But a similar change of constitution, though it eludes common observation, successively takes place in what are deemed perfect fluids, such as water and alcohol. According as the transition from solid to fluid is slow or rapid, it is preceded by softness or brittleness; which may be therefore considered as only the same quality under different aspects. Thus, glass softens by degrees before it melts, and ice turns friable before it dissolves into water.

Such

Such is likewise the case with most of the metals. Tin or lead, for instance, exposed to sufficient heat become brittle, next friable, and then flow down into a liquid mass. The same circumstances will occur, if the fluidity be produced by any other cause. Thus if tin be made successively to imbibe quicksilver, it will first grow brittle, then friable, and next form an amalgam, which will soften more and more, continually approaching to the fluidity of quicksilver itself. In the order of softness, therefore, glass, being remarkably brittle, occupies an intermediate place between the metals and paper or vegetable substances; and such also was found to be its arrangement with respect to the property of emitting heat.

Tinfoil, being formed by passing it between two rollers, must evidently be denser and harder than common tin. This hardness seems to compensate the want of lustre and smoothness of its surface; for its propellent power is equal if not rather superior to that of planished block-tin. On the other hand, block-tin when thus hammered is undoubtedly harder than ordinary tin; and it was lately remarked that the latter has, with respect

spect to the propagation of heat, double the power of the former: an augmentation greater than what can fairly be ascribed to the unevenness of surface, since it is nearly as much as would obtain if this were completely furrowed.

Filling the canister with ice, the impression of the blackened side being as usual reckoned 100, it was increased to 105, by moistening the paint with water; and soft jelly being applied to the clear side, its effect rose to 120. As the difference between the temperature of the canister and that of the room was very small, I would not lay much stress on the accuracy of those numbers; but I can scarcely hesitate to conclude that, in both cases, the energy was sensibly augmented.

I may here notice a fact, which, though of a very different nature, has yet apparently some relation to the preceding. It is well known that, all colours in powder assume a deeper or darker shade when worked up with oil or water. This implies a more copious absorption, at least admission, of the rays of light. Tallow becomes transparent by melting, that is, it receives and does not return the light. But it is perhaps an extension

of

of the same principle that fits a substance to receive the impressions of heat conveyed through the atmosphere. And we have repeatedly observed, that the absorbent and propellent powers are congenerous and equivalent.

5. The last quality, which may perhaps have some influence in modifying the power of a substance to emit or absorb heat, is its *colour*. On colour the disposition to imbibe the rays of light principally depends. Nor is it unreasonable to suppose that this phænomenon is only the result of a more general principle, and that the same constitution of surface which acts on the incident light, may operate, by a more diffusive energy, on the aërial accessions of heat. Black is most absorbent of light, white discharges it most copiously, and scarlet is next in order, its emissions being very bright and dazzling.

## EXPERIMENT XXXI.

I PAINTED three sides of a square tin canister with lamp-black, whiting and minium; each being worked up with as little size as would give consistence

sistence to the pigment. Being presented successively to the reflector, the effect of the black surface was 100 degrees, that of the white only 85, and that of the red one 90 degrees.

These curious facts would seem to countenance the supposition above mentioned. But we must observe that the substances themselves which furnish those colours are very different in their nature. Lamp-black is a species of charcoal and has a vegetable origin, whiting is a fine calcareous earth, and minium is an oxyd of lead. The different constitution of those substances, therefore, independent of the consideration of colour, might afford a sufficient explanation of their various effects. Writing paper applied to the side of the canister has a power very little inferior to that of blacking, being equal to 98 degrees; yet rubbed with chalk, its action is quickly reduced to 90 degrees, or even lower. The colour is still the same, but its energy is thus greatly diminished. Stained paper has very nearly the same action as white paper; and it is only, when coloured by a pigment superinduced, that the diversity of effect becomes conspicuous. There are some species

species even of whites, of a soft glossy quality, that, with regard to the property of discharging heat, surpass lamp-black itself. On the whole, it appears exceedingly doubtful if any influence of that sort can be juſtly ascribed to colour. But the question is incapable of being positively resolved; since no substance can be made to assume different colours, without at the same time changing its internal structure.

CHAP-

## CHAPTER VII.

THE principles deduced in the preceding articles are farther confirmed and extended by experiments to determine the modifications which affect the power of reflection. The power of a surface to reflect heat, as was formerly ſtated, is supplementary to the power of emitting or absorbing it. The one increases, therefore, as the other decreases; the former is greatest, when the latter is least; and if substances were arranged according to their reflecting qualities, the order of their various dispositions to receive or propel heat would be exactly reversed. It was found difficult to distinguish the various aptitudes of metallic surfaces to discharge heat. But the problem can be indirectly solved with peculiar advantage; for the reflective powers of metals being comparatively very great, any differences which may obtain among them will be the more discernible. The most obvious way of proceeding is to have

have similar reflectors constructed of different metals. A mode equally conclusive however, and attended with much less trouble and expence, is to employ a secondary reflection. If a small flat circle, for example, be fixed parallel to the face of the tin reflector and nearer than the proper focus, a second reflection will take place, which, according to the laws of catoptrics, will form a new focus, similar in every respect, and situated as much before the circular plate as the former was behind it.

## EXPERIMENT XXXII.

HAVING procured circular plates of some different metals, as flat and smooth as possible, and five inches in diameter; I planted them successively in the same position, directly facing a large tin reflector, and 5 inches distant from it. The focal length corresponding to the situation of the canister was 7 inches, and consequently the secondary focus was only 3 inches from the reflector. There I placed the differential thermometer, the precise spot in which the action was concentrated being ascertained by help of a lighted taper. The comparative results are exhibited in the following table.

| Reflecting Substance. | Effect. |
|---|---|
| Brass - - - | 100 |
| Silver - - | 90 |
| Tinfoil - - - | 85 |
| Planished block tin - | 80 |
| Steel - - - | 70 |
| Lead - - | 60 |
| Tinfoil softened by the affusion of quicksilver, and with a brilliant surface - | 50 |

A plate of glass, substituted in the place of those metallic ones, produced an effect of about 10. With a coat of wax or oil, the action did not exceed 5.

These few trials exhibit a notable diversity of effects. But the subject might be prosecuted much farther: and I may observe, that it is not necessary that the plates should either be circular or of equal dimensions; because tinfoil can always serve as a standard of comparison, and a coat

coat of it may be applied and removed at pleasure, without affecting the polish of the reflecting surface.

In the experiment just recited, when the second reflector consisted of tin, one third of the force of the canister was lost in the double reflection. That loss ought not to have exceeded one-eighth part of the whole. The deficiency, amounting to one-fifth, must, I presume, be attributed to the sort of aberration formerly remarked.

## EXPERIMENT XXXIII.

DESTROY the polish of a tin reflector by rubbing it with a bit of sand paper in one direction, till the surface becomes completely furrowed. It will now show scarcely the tenth part of its former power in the reflecting of heat. Nor is the focus more diffuse than before, or cast into a more elongated shape; for if the differential thermometer be gradually drawn aside, the impression will still diminish after the same proportion, whether the reflector be placed with its *striæ* parallel or perpendicular to the table.

I need scarcely observe that this experiment might be performed without injuring the reflector, and almost with equal advantage, by having previously coated it with tinfoil, which will adapt itself to the curvature with sufficient exactness. It deserves to be remarked that, if the *striæ* be rubbed across, the power of the reflector will again be somewhat increased; a convincing proof that the diminished action was not caused by the defect of reflection, but by the copious absorption of heat which had preceded that process. It is of the utmost consequence to avoid scratches in cleaning the reflectors. A little whiting answers the purpose very well, only the smallest grain of sand, happening to mix with it, will infallibly mark the smooth surface of the tin. With all the care I could take, when there was repeated occasion to refresh the lustre of a reflector, I found very considerable variations in its effects, amounting in the extreme cases to a full third.

## EXPERIMENT XXXIV.

Let the same reflector, with its striated surface, be directed to the sun. It will form a diffuse image, much elongated; and judging roughly, its power of burning will not amount to the tenth part of its former intensity. This singular want of efficacy proceeds not from any absorption of the light, for the surface appears brighter than before, but from the irregular reflection which prevents the concentration of the solar rays.

Tin reflectors afford likewise the most convenient and satisfactory method of ascertaining the various degrees of influence which different coatings, according to their degree of thickness, are fitted to exert. A few experiments will set this matter in a clear light.

## EXPERIMENT XXXV.

A tin reflector, coated over with a pretty thick layer of tallow, had its power reduced from 100 to 8. But held before a good fire till as much of the tallow was melted off as would flow down, its power was found to be again augmented to 37. Another reflector had its surface covered with a film of olive oil at the temperature of 17 degrees centigrade; its effect was 42, reckoning the same standard as before: being again cleaned and moistened with strong alkaline lye, it produced an impression nearly the same, or 38.

If the tallow, the oil, and alkaline lye in this experiment had formed a coat of considerable thickness, suppose the hundredth or even the five hundredth part of an inch, their action would not have amounted to 10, perhaps not have exceeded 3 degrees. The comparative magnitude of the results was owing evidently to the thinness of those soft films, which admitted the partial action of the metallic substratum. Yet the tenuity was not to an extreme degree, for the iridescent colours had not begun to emerge.

The most eligible mode, however, of exhibiting those varied effects is to employ coatings of isinglass. The procedure requires some little dexterity and more patience. A small portion of liquefied

liquefied jelly is poured into the cavity of the reflector, which then is turned continually round in an inclined position, till the glue has spred equally and congealed over the whole surface.

## EXPERIMENT XXXVI.

To the surface of a large deep reflector, whose entire power was equal to 100 degrees, I applied, in the way just described, a thin layer of liquefied jelly. The initial impression on the focal ball was now 31, and as the coat gradually dried, it rose successively to 44, 48, and 72 degrees, at which last the effect was stationary. A thicker coat being applied, the action at first was only 9 degrees, and afterwards increaſed successively to 13, 16, 19, and 25 degrees. The surface of the reflector being again carefully washed and cleaned, a layer of very thin jelly was spred over it, and the effect on the differential thermometer successively rose from 19, to 32, 40, 58, and at last to 79 degrees, when the isinglass formed a very fine iridescent film.

These facts, compared together, demonstrate

clearly, that the thickness of the coating diminishes the intensity of the reflected heat. But the same principle is finely shown in each single instance, by the progressive augmentation of effect which takes place during the progress of drying; for the coat of jelly must obviously grow continually thinner, till it has at last acquired the solid consistence of parchment. When the jelly is applied sufficiently dilute, every gradation almost is successively exhibited, from the feeblest reflection to that of tin itself. But it is of moment towards discovering the nature of that influence, to determine the degrees of reflective power corresponding to the various measures of tenuity of the superficial crust.

## EXPERIMENT XXXVII.

I was enabled, in the manner already described, to apply coatings of any required thinness to the cavity of the tin reflector; for narrow delicate slips of dried jelly were cut to the calculated dimensions, and redissolved in boiling water. Reckoning,

koning, as usual, the entire effect of the reflector to be 100 degrees, it was thus regularly diminished by the successive coatings.

| Thickness of coating in parts of an inch. | Effect of Reflection. |
|---|---|
| $\frac{1}{20,000}$ | 77 |
| $\frac{1}{10,000}$ | 49 |
| $\frac{1}{5,000}$ | 37 |
| $\frac{1}{2,000}$ | 27 |
| $\frac{1}{1,000}$ | 19 |

It is impossible to spread a coat strictly of uniform thickness over a round surface. I will not therefore pretend to state those numbers as rigorously exact: I would consider them rather as approximations to the truth, but sufficient, however, to give us notions much more precise than before.

In pushing the experiment farther, a phænomenon occurred which struck me with some surprise. Having coated the reflector with a film of isinglass the 400,000th part of an inch thick, I expected of course an effect somewhat below 100 degrees. But it was actually 125 or one fourth

fourth part greater than the simple power of the reflector before such coating was applied. It soon occurred to me, however, that the power of the same reflector is liable to considerable variation by having its surface refreshed, and that, having repeated occasion to brighten up the one which I then used, its original polish was hence probably much impaired. The application of liquefied jelly would therefore fill up the scratches, and, in effect, restore the surface to its former smoothness. The action of the coated reflector would almost be the same, as if it had previously received its highest lustre and finish. This conjecture seems to be confirmed by other observations; for as the thickness of the coat is increased, the corresponding power of reflection still uniformly diminishes. Thus the effect when a coat of the 700,000th part was applied, was 118 degrees, and that corresponding to a coat of the 500,000th of an inch, only 110.

Supposing the entire original effect of the reflector to be 127 degrees, the comparative effects produced by the application of coatings of different thickness will stand thus.

Thickness

| Thickness of the coating in parts of an inch. | Effect. |
|---|---|
| $\frac{1}{400,000}$ | 98 |
| $\frac{1}{100,000}$ | 93 |
| $\frac{1}{50,000}$ | 87 |
| $\frac{1}{20,000}$ | 61 |
| $\frac{1}{10,000}$ | 39 |
| $\frac{1}{5,000}$ | 29 |
| $\frac{1}{2,000}$ | 21 |
| $\frac{1}{1,000}$ | 15 |

If we presume that the reflective power of isinglass is equal to 8, the influence of the proximity of the metallic surface will be exhibited in this table.

| Thickness of the film in millionth parts of an inch. | Influence of the substratum of tin. |
|---|---|
| $2\frac{1}{2}$ | 90 |
| 10 | 85 |
| 20 | 79 |
| 50 | 53 |
| 100 | 31 |
| 200 | 21 |
| 500 | 13 |
| 1000 | 7 |

It is not easy to perceive the law of progression. When the coating is of considerable thickness, the influence of the metallic substratum is nearly

nearly in the inverse ratio of that thickness. But when it forms only a minute film, the variation of effect is much slower. The most obvious way of representing the progression, is to suppose the thickness of the coating to be affected by some constant quantity. Thus, if we assume, what appears not improbable, that the centre of action of the tin is situate at the $\frac{1}{1250}$ of an inch, or 80 millionth parts below its surface, and likewise admit that the influence exerted is reciprocally as the distance of that point from the external surface of the coating; we shall obtain, with tolerable nearness, the numbers above stated.

| Thickness of the film in millionth parts of an inch. | Calculated influence of the substratum of tin. |
|---|---|
| $2\frac{1}{2}$ | $89 = \frac{80}{82,5} \times 92$ |
| 10 | $82 = \frac{80}{90} \times 92$ |
| 20 | $74 = \frac{80}{100} \times 92$ |
| 50 | $57 = \frac{80}{130} \times 92$ |
| 100 | $41 = \frac{80}{180} \times 92$ |
| 200 | $26 = \frac{80}{280} \times 92$ |
| 500 | $13 = \frac{80}{580} \times 92$ |
| 1000 | $7 = \frac{80}{1080} \times 92.$ |

Hitherto

Hitherto we have examined only the case where a film of glue is spred on a metallic surface, that is, where the external coat is by its constitution the most fitted to receive or discharge heat. It is much more difficult to reverse the experiment, and to apply a fine metallic plate which is the least receptive of heat, to the surface of glass or of metal. This mode of procedure, however, exhibits a remarkable contrast of effect, very important towards unfolding the hidden springs of action. I employed gold leaf, and foil of silver and copper. By weighing a square of each, I computed the thickness of the gold-leaf to be about the 300,000th part of an inch, that of the silver-leaf, the 150,000th part of an inch, and that of the copper or brass leaf, the 50,000th part of an inch. In applying these to the surface of glass, and especially of metal, it was requisite to avoid using size, which might affect the results. The way which I found answer, was to breathe upon the surface intended to be gilt, and then press it gently against the metallic leaf, which will continue to adhere even after the humidity has evaporated.

EXPERIMENT

## EXPERIMENT XXXVIII.

I WAS fortunate enough to procure a canister four inches square, formed of planished block tin, except one side which consisted of glass. The whole was perfectly tight, and capable of holding boiling water. The glass was a sort of mirror, with a solid plate of pewter applied to the back, instead of the usual amalgam of quicksilver and tin-foil; which pewter, having been poured melting hot, united firmly to the glass, and served as a medium for soldering it to the metal.—The apparatus being properly disposed, and the glass side presented, successively gilt with gold, or silver, or copper; the impression on the focal ball was very nearly the same as that made by a bright surface of tin. The difference was scarcely one tenth part, either in excess or defect; and, with such minute quantities, it were idle to pretend to greater precision.

Thus, notwithstanding the extreme tenuity of those metallic leaves, the action of heat is the same, or nearly the same, as if they had a considerable

siderable thickness of substance. No visible influence is exerted by the vitreous substratum. Yet with films of isinglass attenuated to an equal degree, the interior surface displayed almost its full effect. If similar energies, therefore, had in this instance likewise obtained, the impression made by the gilded surface of glass, instead of being only 10 or 12 degrees, would have ranged between 90 and 100.

## EXPERIMENT XXXIX.

Gild the bright tin sides of the canister with gold, silver, and copper, but without using any size. The action now produced on the focal ball is almost the same as when the mere surface of tin was presented to the reflector.

## EXPERIMENT XL.

Gild in the same manner the focal ball itself, and dispose the apparatus as usual. The coloured liquor will in every case rise to about the fifth part only of the height to which it mounted

when

when the naked ball was exposed; nor does the addition of one or more coats of metallic leaf occasion any sensible difference in the effect, which is almost the same as when a covering of tinfol was applied.

From these concurring facts, therefore, we may safely conclude, that the tenuity of gold, or silver, or copper, leaf, such at least as the shops afford, has no perceptible influence in modifying the discharge or absorption of heat. This tenuity was incomparably greater than what, in the case of isinglass, had begun to affect so palpably the results. But it would be hasty to infer that the thickness of the metal is altogether incapable of altering the phænomena. On the contrary, every thing leads us to presume, that, if the metallic leaves were attenuated in a higher degree, the peculiar effects would be developed. The most likely mode of experimenting would perhaps be to have delicate gold or silver enamel laid on a surface of glass. And this experiment I have since performed. The focal ball incrusted with a fine enamel, was found to exhibit about double the effect that takes place when

when it is coated with gold-leaf. On examining the enamel, with a microscope, it presented an irregular film, with numerous interstices; but it evidently covered by far the greater part of the vitreous surface.

The diversified effects which we have stated are, therefore, incontrovertibly dependant on the degree of the thickness of the superficial film. But this thickness may be viewed as the measure, either of the substance of the coating, or of the distance interposed between the external surface and its heterogeneous substratum. It is consequently still a question, whether those effects are derived immediately from the action of the attenuated film, or result from some modifying influence exerted by the proximate layer. And, in a matter of such extreme subtlety, it is hardly possible to give a direct solution. If it were practicable to have a fine sheet of talc held parallel to the polished metallic side of a canister and capable of being adjusted to different minute intervals by help of a micrometer screw, we should obtain a clear answer to the question. There is a simple fact, however, which fortunately enables us to decide with tolerable certainty. If a piece of gold-beater's skin, of sufficient dimensions, be stretched about an inch before the blackened side of the canister, the diminished effect is the same as if a sheet of paper or parchment had been substituted in its stead. But it has been already proved that the action of screens results from the combined operation of their receptive and propellent properties. These properties must therefore obtain, to the usual degree, in a pellicle whose thickness exceeds not the 3,000th part of an inch. Yet so thin a pellicle, if spread over a surface of metal, would manifest, with respect to heat, very different affections from those of a thick crust of similar substance. Its peculiar degree of energy must, consequently, be referred to the proximity of the metallic substratum. The direct action of the external surface is thus mingled with the force derived from the internal layer, and which augments as the distance from its source diminishes.

CHAP.

## CHAPTER VIII.

WE have thus deduced a train of phænomena which must be deemed equally novel and striking. Our next business is to discover what principle will connect together those curious facts. But before we attempt that investigation it will be expedient to ascend a little higher, and inquire into the constitution of the external world.

All bodies may be considered under the two-fold view of form and substance; the former being mutable, the latter permanent. This distinction is of the remotest antiquity, and has proved the source of much idle refinement in the schools. But it is not, on that account, the less real or important. The whole experience of life displays only a fleeting succession of changes. Nothing seems really stable or quiescent. The scenes are perpetually shifting, and the objects around us are ever presenting new aspects. And what an astonishing variety of forms is even the simplest body capable of assuming! How different in their appearance are the meteors of rain, and hail, and snow? What a wide contrast between water and solid ice on the one hand, and subtle invisible steam on the other? By the agency of intense heat, the hardest substance will melt and rise into vapour. A bright ductile piece of metal passes successively into an earthy oxyd, and a pellucid glass. The diamond itself is of the same nature as charcoal; yet what a vast interval apparently between such a rude material, and the hardness, and dazzling lustre of that precious gem? The changes which can be produced in the composition of bodies by the play of chemical affinities are most various and extensive. But the plastic powers displayed in the process of vegetation and animal life, infinitely surpass the resources of art. Many plants are fed by water and air alone; and in general, the soil performs only a secondary and subordinate office. Those fluids, therefore, which were once esteemed elements, are capable of being transformed into all the diversified products of the animal kingdom; into charcoal, earths, salts, oils, gums, nay, the oxyd of iron,

iron, and perhaps those of other metals. Still more varied are the animal products. In short, we cannot reasonably doubt that every substance is convertible into every other. Nothing seems more chimerical than to indulge a hope, that the human powers will ever be able to achieve the transmutation of earths into the precious metals: but those hapless visionaries who consumed their days in the obscure search after the philosopher's stone, did not, like their fellow labourers who sought the perpetual motion, advance pretensions which involve a physical absurdity. The peculiar properties of bodies must result merely from the different arrangement and configuration of their integrant parts. The substance is in all of them essentially the same; and the sublime scene of the universe owes all its magnificence and splendour to the variety of its composition. We behold a system of perpetual fluctuation; the materials remain, indeed, unaltered, but nature toils inceſſantly to demolish and to renew her stupendous fabric.

Amidst all the various changes that bodies are capable of undergoing, there is one property, and one only, which remains unalterably the same: It is their *weight*, or their gravitation towards the mass of the earth. This force is the aggregate of the single forces exerted by each component particle, and consequently depends merely on the quantity of matter contained. But the virtue of attraction extends indefinitely through space, and constitutes an original and absolute principle of nature. Its intensity is found to decrease as the square of the distance; a law of admirable simplicity, the application of which to the celestial phænomena, seconded by the aids of the higher geometry, has formed the most perfect and beautiful of all the sciences. That law, however, cannot obtain universally; for, if the mutual action of the particles of matter were ultimately attractive, bodies brought near each other would necessarily, like the globules of quicksilver, agglomerate into spherical masses, exactly regular and homogeneous, and of the greatest possible density. Within a certain limit, therefore, the principle of gravitation does cease, and is superseded by other dispositions; or we must admit that it is a branch only of an universal system, which, pervading the whole

whole compass of nature, connects and upholds the material world. The latter hypothesis is more agreable to analogy, and unveils a sublime prospect. The mutual action of the particles of matter is evidently dependant on their distance; but, instead of supposing this force to follow strictly the inverse duplicate ratio of the distance, we may presume that it bears a more general and complex relation, or that, in the language of the modern analysts, it is a *function* of the distance. At distances very remote, the corresponding force will insensibly coincide with the ordinary law of gravitation, which determines the figures and regulates the motions of the planets. But when particles come to be situate near each other, their action must, at a certain point, become negative, or change from attractive to repulsive: and alternate transitions, from the one state to the other, must repeatedly occur within very narrow limits; as appears decisively, from the affections of light in passing by the edges of bodies, and from the beautiful phænomena of thin iridescent plates. Still nearer its source, however, the power exerted must be invariably repulsive, augmenting rapidly, in proportion to the degree of proximity, else matter would permanently collapse. To assist the imagination, the *function* of material action may be represented by an indefinite curve line, of which the abscissæ mark the distances, and their ordinates denote the corresponding forces. Both the final and the initial branches of the curve will be clearly assymptotic, the former approaching continually the extended axis, and the latter bending along a perpendicular drawn downwards from its origin. Where the curve repeatedly crosses the axis, are so many quiescent points, in any of which a particle being placed will continue in equilibrio. But this equilibrium is of two kinds, the stable or the instable, the former easily recovering itself from any slight displacement, and the latter, when once disturbed, being irremediably dissolved; similar to what obtains in mechanics, according as the centre of gravity lies either directly above, or directly below, the point of suspension. If the curve in its progress crosses the axis from the side of repulsion to that of attraction, its intersection will evidently be a point of stability; for if a particle be pushed inwards,

it

it will then be repelled back again; and if it be pulled outwards, it will experience an attractive force, which will recall it to its first position. And such appears to be the most general constitution of bodies. They are all susceptible of contraction and dilatation, and, for the most part, they make an effort to recover from the impression of external violence. Nay, this reaction is exactly equivalent to the external pressure, and every substance, even the softest, is within certain limits perfectly elastic. Thus, water placed in a condensing engine, and consequently compressed equally on all sides, will, on the force being removed, regain accurately its former bulk. And, if water be suddenly struck, so as not to allow time for the change of figure or the recession of the adjacent parts, it will manifest all the resiliency of the hardest elastic body; which is clearly evinced in the familiar play of duck and drake.

When the curve passes from attraction to repulsion, its intersection with the axis is a point of instable equilibrium; for, in proportion as a particle is pressed inwards, it will be pulled forcibly rom its position; and if it be drawn outwards, the

the repulsion, now conspiring, will bear it along with accumulating power. Such, perhaps, is the intimate structure of some fugitive gases.—It is not necessary, however, to suppose that the curve of elemental action is always continuous; it may suddenly break off, and leap to the other side of the axis. Mathematicians, indeed, have laid great stress on what they style, *the law of continuity*; but since the controversy between Euler and Dalembert concerning the vibrations of the musical string, that opinion or prejudice is likely soon to lose, in some degree, its influence.

However paradoxical it may seem, it is more a question of refined curiosity than real importance, to inquire, if the ultimate particles of matter have any positive magnitude, or are mere mathematical points. The arguments commonly brought to demonstrate the infinite divisibility of matter are perfectly nugatory. With all their furniture of diagrams, they prove or illustrate nothing more than a very simple conception of the mind, namely, that there is no portion of matter, however small, but we can imagine one still smaller. It is preposterous, however, to judge

the

the constitution of the universe by such abstractions. The term infinite, though darkened by mysticism, conveys merely a negative idea; it excludes whatever is limited and defined; it supposes the powers of the mind not to repose, but to strain continually to rise to a higher pitch. In the physical world, on the contrary, every thing appears to be individual and determinate. Though experience informs us to what astonishing degree matter can be attenuated, it is most congruous to believe that there are certain fixed or impassible limits at which the capability of farther subdivision utterly ceases. Those ultimate corpuscles are therefore the eternal atoms of Democritus, or the sentient monads of Leibnitz. On this hypothesis, however, the primordial line of action is a physical and not a mathematical, curve; or it is not strictly curved at every point, but proceeds by successive minute gradations, corresponding to the breadth of the elementary particles. It is even from the consideration of such serrated lines, that we derive the theory of curves: we strive unremittingly to multiply the sides of the polygon and smooth away its asperities;

ties; we descry boundaries to which we are continually tending. In like manner are grounded the principles of the differential calculus. We survey a series of quantities with sensible differences; we perpetually diminish those differences, and seek the relation to which this procedure points. But in the physical curve or polygon, there are certain limits in the process of exhaustion, on which the imagination would ultimately rest.

All bodies, thus, consist of physical points, endued with certain powers, attractive or repulsive, which repeatedly interchange and vary their intensity with the distance, according to some uniform and permanent law. This universal law of action, constitutes the essence of matter; it is original, absolute, and underived; and the numerous properties of corporeal substances, with all their apparent diversity, are only the several results or developements of the same grand principle. Could we discover the equation of the primordial curve in its whole extent, we might thence unfold the internal structure of bodies, deduce their respective qualities, and estimate the effects of chemical agency,

agency, with the same rigorous accuracy as we now calculate the planetary motions. It were too sanguine, however, to expect that science will ever make such mighty advances. The initial part of the curve of action must evidently be more intricate than its remote branch, and we are besides, precluded almost totally from the means of ascertaining its nature. The microscopic world seems to retire from human research, and to conceal itself in impenetrable obscurity. On a subject so very recondite, we shall probably, for ages, have only vague and inadequate ideas.*

But though the resolution of bodies must ultimately terminate in atoms, there may be distinct stages in the progress of the subdivision. What is usually denominated a particle, for instance, may be regarded as a cluster of atoms, arranged in some of their positions of stability; and the action of such a particle will be the result of the compound forces directed to all the points of this secondary system. A series of successive compositions seems to disclose itself in the internal structure of crystals.

* See Note XI.

The

The preceding views lead to a curious result, which, though it may appear merely speculative, will be found, in the explication of a variety of phænomena, to be of real and extensive importance. It is this—that the communication of motion is not strictly instantaneous, but requires some finite portion of time. If, for instance, I push the end of a long rod in the direction of its length, the remote extremity does not simultaneously advance; or if one ball impinges against another, the force is not transferred at the very moment of collision. Consider a string of connected particles: the first is impelled towards the second; it therefore makes a certain approach, till the shock is extinguished by the accumulated powers of repulsion; but in this constrained position of proximity to the second particle, the first repels it, and causes it to make a similar approach to the third. And thus, by successive partial vibrations, the original impulse is transferred along the whole chain of particles. This intestine process may be rendered more familiar to the imagination, by examining what takes place when a stroke is propagated through a spiral or helical spring. If I give a twitch near the

the one end of a long cord stretched tight, the jerk, forming a slight sinuosity, will visibly dart along the line. In ordinary cases of impact, the time elapsed is so extremely small as to escape the most attentive observation. Its effects, however, are not the less manifest. If any ivory ball strikes against another of equal weight, there should, according to the common theory, be an exact transfer of motion. But if the velocity of the impinging ball be very considerable, so far from stopping suddenly, it will recoil back again with the same force, while the ball which is struck will remain at rest. The reason is, that the shock is so momentary as not to permit the communication of impulse to the whole mass of the second ball; a small spot only is affected, and the consequence is therefore the same as if the ball had impinged against an immoveable wall. On a perfect acquaintance with such facts and their modifications, depends in a great measure the skill of the billiard-player. It is on a similar principle, that a bullet fired against a door which hangs freely on its hinges, will perforate without agitating it in the least. Nay, a pellet of clay, a bit of tallow,

tallow, or even a small bag of water, discharged from a pistol, will produce the same effect. In all these instances, the impression of the stroke is confined to a single spot, and no sufficient time is allowed for diffusing its action over the extent of the door. If a large stone be thrown with equal momentum and consequently smaller velocity, the effect will be totally reversed: the door will turn on its hinges, and yet scarcely a dent will be made on its surface. Hence likewise the theory of most of the tools and their mode of application in the mechanical arts: the chissel, the saw, the file, the scythe, the hedge-bill, &c. In the process of cutting, the object is to concentrate the force in a very narrow space, and this is effected by giving the instrument a rapid motion. Hence too the reason why only a small hammer is used in rivetting, and why a mallet is preferred for driving wedges. But I must check this digression. The idea has been almost entirely overlooked by writers on mechanics. I content myself at present with throwing out these few hints: were the subject pursued through all its branches, with that copious illustration which it merits, it would

would produce a very interesting work. It is in certain cases somewhat ridiculous, to see philosophers pretending to instruct artists: they ought first to become disciples; for the workshops may be considered as the great schools of experiment, and still afford the best lessons for improving and refining science.

Not only is a sensible time required for the propagation of motion, but it is possible, in every case, to calculate the actual celerity of its transmission. Since the impulse is conveyed along the particles by a sort of pulsation, the nature and periods of those pulses can be determined by the received principles of dynamics. In the same substance, all the vibrations, whether great or small, must be evidently isochronous; in different substances the rapidity of their succession will depend on the joint but opposite consideration of density and elasticity. The duration of each pulse will be directly as the interval between two adjacent particles, and inversely as the square root of the height of the column which measures the elasticity; the celerity of communication is therefore in the direct subduplicate ratio of the height of that homogeneous column. Impulse is thus transmitted through air, precisely in the same manner, and with the same velocity, as sound. In the other gases, the rate of its propagation is easily inferred; in the hydrogenous, for example, motion must travel with rather more than three times the ordinary swiftness of sound. With regard to liquids, their elastic power, or the degree of compression required to produce a certain contraction, must be previously ascertained. This, in a very few cases, has been determined by the nice measurement of the effect of atmospheric pressure. It hence follows, that an impulse will be transmitted through water, with about $3\frac{1}{2}$ times the usual velocity of sound.* To estimate the expansive force of solid substances is a more difficult point. I have, however, succeeded by an indirect mode. If a long bar be supported at the two ends in a horizontal position, it will swag or bend downwards, forming a curve which in small deflexions may be considered as a portion of a circle. The upper filaments, belonging to a smaller circle, will be contracted, and the

* See Note XII.

under

under ones, for a contrary reason, as much extended. The condensation produced on the one side, and the distension on the other, by the lateral action of gravity, are the same as if two forces had been applied longitudinally, each equal to half the weight of the bar when prolonged to the length of the radius of curvature. Hence I conclude, that motion is conveyed through deal with $5\frac{1}{2}$ times the velocity of sound, through soft iron with only twice that velocity, and through hard hammered iron with five times the same velocity.*

It thus follows, that sixteen hours will elapse before an impulsion is propagated through the atmosphere, from pole to pole. A similar transmission would be performed through the superficial waters of the ocean, within the space of four hours and thirty-five minutes. The rapidity in either case incomparably surpasses that of the fleetest wind; and hence the reason of the sighing in the air, and the murmuring on the shore, which are commonly believed to betoken a distant but impending storm.† Perhaps it is from

* See Note XIII. † See Note XIV.

the progressive communication of impulse through the atmosphere, that we are to seek the true explanation of the variations of the barometer, which have so long perplexed philosophers. Wind is, no doubt, the primary cause of those fluctuations, by disturbing the equilibrium of the air. But, submitted to rigorous calculation, such action is found quite inadequate to produce the effect: the most violent hurricane, according to Dalembert,* ought not to occasion a descent of the mercury equal to the tenth part even of what is frequently observed. I must remark, however, that those investigations are founded on the erroneous supposition, that each varying impulse is *instantaneously* diffused over the whole surface of the globe. In reality, the place where the tornado originates, may be considered as in a manner insulated or detached from the remote regions of the asmosphere, since many hours must elapse before they can feel or endeavour to poise its influence. Whatever inequalities therefore might arise, they will evidently be much

* *Essai sur les Vents*, which was crowned by the academy of Berlin.

augmented

augmented from the effect of confinement and concentration; and it would be an object of curious, though most difficult investigation, to determine with precision how far the consideration now mentioned is capable of modifying and heightening the simple results.*

All forces are radically of the same kind, and the distinction of them into *living* and *dead* is not grounded on just principles. It is a fundamental theorem in dynamics, that a body has the square of its velocity increased or diminished in the ratio compounded of the accelerating or retarding pressure, and the space in which this operates. The same velocity, or momentum, will therefore be generated, if we augment the pressure, and diminish proportionally the extent of action. The mutual exchange may be pushed so far, that the little space of orgasm shall become altogether insensible. It is this extreme case which seems to have given occasion to misconception and incorrect ideas. A tennis-ball dropt from some height, will acquire a very considerable velocity in its fall; thrown from the hand, it will receive the

* See Note XV.

same velocity in a much shorter sweep; but the arm exerts a pressure much greater than the weight of the ball, or the simple force of gravity: struck with a racket, the same motion will be communicated within a space scarcely discernible; and if the ball so soon leaves the racket, it is likewise urged by a force proportionably great. A blow and a squeeze, however different in their effects, are strictly of the same nature. In the one the force is partial and concentrated, in the other it is general and diffuse.

The whole scope of this reasoning, it will be perceived, is in direct contradiction to the noted axiom of the schools, that "nothing acts where "it is not." But I would observe, that all axioms are merely the simple conclusions, drawn *à posteriori*, from familiar experience; that however fitted for the ordinary business of life, they are useless and even prejudicial in philosophy; and, that being derived from loose and superficial views, they often require restriction, and are liable to inaccuracy. In matters of science, the general opinion of mankind, termed common

sense,

sense, is always a very suspicious standard of appeal. If a body acted only within itself, it is clear that the force could never be transmitted; there would be no communication, no sympathy with the rest of the universe. In vain shall we have recourse to the agency of invisible *intermedia:* the interposing of successive stations may divide, but will not annihilate distance; and, after torturing our imagination, the same difficulty still recoils upon us.

It is a remarkable and instructive fact in the history of philosophy, that impulsion should have been at one period the only force that was admitted. The motion of a falling stone was certainly not less familiar to the senses than that of a stone which is thrown; but in the latter case, the contact of the hand was observed to precede the flight of the projectile, and this circumstance seemed to fill up the void and satisfy the imagination. Gravitation sounded like an occult quality; it was necessary to assign some mechanical cause; and if there were no visible impulses to account for the weight of a body, might not that office be performed by some subtile invisible

 agent?

agent? Such was the sway of metaphysical prejudice, that even Newton, forgetting his usual caution, suffered himself to be borne along. In an evil hour he threw out those hasty conjectures concerning æther, which have since proved so alluring to superficial thinkers, and which have in a very sensible degree impeded the progress of genuine science. So far from resolving weight or pressure into impulse, we have seen that the very reverse takes place, and that impulse itself is only a modification of pressure. This statement has already some distinguished adherents, and must in time become the received opinion. Science has experienced much obstruction from the mysterious notions long entertained concerning causation.*

* See Note XVI.

CHAP-

## CHAPTER IX.

I HAVE dwelt so long on these preliminary disquisitions, that I may well seem to have lost sight of the principal object. But I have seized the occasion to set before my readers a variety of physical views, some of which are new, and all of them remote from common apprehension. Notwithstanding their apparent subtlety, however, they will, I trust, be found on reflection to be upon the whole solid and important. They will facilitate our progress, render our future reasonings more intelligible, and enable me to abridge the thread of investigation. I now return to consider the Nature of Heat.

It is almost superfluous to remark, that the term *heat* is of ambiguous import, denoting either a certain sensation, or the external cause which excites it. This last sense only we have to consider. Our feelings furnish a most imperfect notice of the measure of heat: indicating merely the

the impression made upon the human organs, they depend on the relative temperature and condition of our body, the quickness or slowness of communication, the particular quality of the contact, and a variety of other circumstances. A substance of the same warmth as the hand, feels agreeable; as it grows continually hotter, the corresponding sensation grows more intense; and at length becomes absolutely painful, and would terminate in the destruction of the nervous fibrils. Suppose the procedure to be reversed, by constantly abstracting the heat, or, in common language, exposing the substance to cool; the impressions of touch will then be throughout of the same kind. The feeling is at first only unpleasant, then grows more and more painful, till the sentient organ at last becomes numb, and is destroyed. The extreme sensations are thus in both cases analogous; for it is a curious, though not a very consolatory fact in the animal œconomy, that all our feelings, whether of pleasure or of pain, when pushed to their utmost extent, are invariably painful, and the transition is often sudden, from exquisite delight to the most acute torture.

The

The epithets of hot and cold are, therefore, merely relative, and mark the excess or defect of temperature in estimating from a common standard. But this standard is evidently accidental and arbitrary; and it is expedient in philosophical discourse, to assume or imagine a point of reference placed much lower, so as to include every possible condition of substances in the same positive range. Cold is thus, correctly speaking, only an inferior degree of heat. The ascending scale of temperature seems almost boundless; the descent is difficult, and confined within narrow limits. With the actual production of heat we are familiarly acquainted. The contrary process is practicable only to a small extent, and in a few solitary instances.

It has long been disputed, whether heat is simply a state or condition of which all bodies are susceptible, or results from the infusion of some distinct and active principle. In the infancy of science it was supposed to consist in certain intestine vibrations; and this opinion, however vague and undefined, has still some adherents. It promises, indeed, at first glance, to satisfy the imagination,

imagination. Fire itself is generated by friction, and no association is more natural than heat and motion. But the shapeless hypothesis will not bear a strict examination. In reality, it explains nothing; it throws out a delusive gleam, and then leaves us in tenfold darkness. Does the whole of the heated mass simultaneously vibrate by alternate contraction and dilatation? Or, are these only numerous partial oscillations, diffuse but concatinated, emerging in quick succession? And yet, on either supposition, what precise idea can we annex to the *degree* of heat? Is it determined by the magnitude, the frequency, or the force of the pulses? We are thus led into a labyrinth of perplexities. But the opinion, that heat consists in vibrations, is not merely nugatory; it is exposed to insurmountable objections. And not to multiply words, I shall confine myself to two capital points: 1. The communication of heat from one body to another, would require such a rare concurrence of circumstances, that it could very seldom obtain in nature. For the vibrations excited by contact, must either be strictly in unison with the first, or must form some simple concord:

cord: but the interval of the vibrations will clearly depend on the extent of the affected substance, and its degree of elasticity. In two substances of the same kind, if the one bears no elementary ratio to the other, it cannot sympathize with the primary pulsations. Nay, without such a singular coincidence, those pulsations, so far from communicating their influence, would themselves be extinguished. This is what we experience in sonorous bodies. When one in a state of vibration is made to touch another not fitted to yield the same, or at least concordant notes, a sudden jarring ensues, and the sound quickly expires. 2. But in opposition to the opinion that heat consists merely in certain intestine motions, I may urge another objection equally conclusive, and more easily comprehended. Admitting that hypothesis to be real, all heat must gradually relax and die away: for this is the fate of every species of motion experienced upon earth; and with respect to intestine motions in particular, they suffer such manifold obstructions, and are attended with such waste of power, that they speedily terminate. But in every case where a body

body is observed to cool. it only distributes its heat among the surrounding matter; and the loss sustained on the one part, is exactly compensated by the accession made on the other. If the communication be rendered more imperfect, the exchange or transfer of heat will become proportionally slower. Were it possible to procure an absolute vacuum, a body thus insulated would indisputably retain for ever the same temperature.

Heat is universally accompanied with expansion. To this law there is no real exception. In the progress of heating indeed, a body may totally change its form or internal arrangement, which may occasion indifferently either dilatation or contraction; but after such change of constitution is effected, the subsequent rise of temperature is regularly marked by a corresponding distension. Ice, so long as it retains that character, expands by the accession of heat, as well as in the state of water.

And what causes this expansion? The particles of a heated body recede from each other; a certain repulsive force is therefore introduced; but whence

whence does it actually proceed? To suppose that matter can of itself assume or acquire new energies, is inconsistent with the uniform testimony of experience, and most repugnant to the whole train of our ideas. It is true, that in magnetism and electricity, neither of which, perhaps, demands the existence of a peculiar fluid, cases occur where substances are made to exert mutual attractions and repulsions, only in consequence of the approximation, or mere apposition of some distinct body in the due state of excitement. But this superinduced action is fleeting and dependant; no permanent character is imprinted, and the sympathetic substances relapse into their ordinary state, the moment the excited body is withdrawn. The phænomena of heat are totally of a different nature. There is no solid objection to the materiality of heat, because it is not visibly present in bodies. The existence of many things we must infer from their undoubted effects. Water, for example, is lodged concealed in the substance of apparently dry wood, from which, by violent compression, it may be squeezed out in its liquid form.

The

The expansion therefore which accompanies and indicates heat, is produced by the infusion of some extraneous matter. This must be absorbed by a general attraction into the substance of the body affected; for simple aggregation, or mechanical addition, could not develope any active efforts. It must likewise be a species of fluid, since fluidity is a condition indispensable towards every chemical union. But it may be still of a liquid, or of a gaseous nature; and there are corresponding, two modes in which it would produce distension, either directly by repelling the proximate particles of the body with which it is combined, or indirectly by exerting repulsion among its own particles. 1. Though the igneous fluid attracts all substances considered in the mass, it is nowise inconsistent to admit, nay, the primordial system requires, that within the limits of contact, this action shall change into repulsive. Analogous instances are numerous. Dry wood, it is known, soaked in water, swells, and with astonishing force. The humidity insinuates itself between the fibres and distends them. If copper be made to imbibe quicksilver, it will likewise expand.

expand. These are strictly both cases of chemical combination, but in which the solid predominates. They exhibit a real concentration of matter, for it can be clearly proved, that the bulk of the compound is invariably less than that of the two component parts taken separately. A sort of intermediate character is likewise acquired. The wood shows its participation of fluidity by its increased softness: and the incipient passage of the copper to quicksilver is marked by the whiteness, and still more by the brittleness, which are induced. If the basis to which the liquid is joined be liquid also, a similar condensation of matter will yet take place. This we witness when a small portion of sulphuric acid is poured into water. 2. The other mode in which the igneous fluid might cause expansion, is by introducing the influence of a repulsive force subsisting among its own particles. This implies that its separate existence is of a gaseous or aëriform nature. The particles of the fluid, being dispersed through the heated mass, and powerfully attracted by it at the same time that they repel each other, must in mutually receding produce a general dilatation. As

 the

the gaseous fluids, in comparison with liquids, have a much wider sphere of action, so the attraction to their bases is proportionally more extended. That attraction will therefore be scarcely at all affected by any partial repulsion which might arise from too close proximity. The action of an expansive fluid incorporated with a liquid or solid substance, is less familiar than the former case; yet there are not wanting facts to confirm our reasoning. Water, in its ordinary state, contains a notable portion of common air. This is rendered obvious by the affusion of any strong acid, which immediately expels a stream of minute bubbles. The want of air constitutes indeed the sole difference between fresh distilled water, and that which, having been for some time exposed to the atmosphere, has re-absorbed that fugacious element. This union however is not very powerful. The imprisoned air is disengaged from water in a variety of processes: by freezing—by boiling—introduced above a long column of mercury—or under the receiver of an air-pump. Such is its feeble adhesion to the basis, that it recovers its elastic form the moment the incumbent

incumbent pressure of the atmosphere is removed; and, for a similar reason, the application of heat, by distending it, produces a partial separation. Water necessarily discharges its air previous to the act of congelation; a circumstance which sometimes retards that species of crystallization. Hence it is, that the water freezes sooner which has been boiled. If having filled two wine-glasses, the one with crude, the other with boiled, water, I expose them to a sharp cold; the former will present a cake of ice crowded with large bubbles entangled in the mass; the latter will contain a solid lump, almost pellucid, with only some minute specks or striæ shooting from the centre. Water placed within a receiver which is partly exhausted, will scarcely freeze at all, and will only form a loose spongy concretion; for the globules of air which are successively extricated in the process becoming unusually rarefied, and of course less apt to escape from the clustering icicles, occasion an excessive swelling. Ice frozen in large masses, and therefore slowly, is always of the most solid and uniform texture. Time is allowed for the latent air to disengage itself.

itself. The external portion, too, of ice which is the first formed, is likewise the clearest, for the air globules are constantly invaded and driven inwards.

But the expansion is more apparent which the presence of a gaseous fluid communicates to solid bodies. Thus lead, by absorbing pure air, is converted into minium, and with that change has its density greatly diminished, and consequently its bulk enlarged. It will be perhaps urged, that the effect is produced by oxygene, and not the oxygenous gas. Much as I admire the general simplicity and elegance of pneumatic chemistry, I cannot admit the accuracy of some of its tenets. Pure and inflammable airs are real and intelligible substances; but are not oxygene and hydrogene, their supposed bases, like the defunct phlogiston, only mere fictions of system? Their existence is inferred from a strained and inverted analogy. Lead acquires only one-tenth part more weight in being converted into minium, and this small accession of matter, whatever was its density, could not by its passive dispersion introduce such a very considerable augmentation of volume. This betrays

trays repulsions which could belong only to a gaseous fluid. And if the mutual repulsions among the particles be made to exceed their adhesion to the metal, the fluid will again recover its elastic form. The simple application of heat in some instances, and the assistance of conspiring affinities in others, will disengage the oxygenous gas from its metallic basis. Nor is it improbable, that in certain cases the mere removing of the atmospheric pressure would in part produce the same effect.

The question now recurs, to which species of fluid—the liquid, or the gaseous—must heat be referred? Bodies, in being heated, acquire no sensible increase of weight; but their corresponding expansion is often very considerable. The mere insertion of such a minute portion of matter is therefore altogether inadequate to the production of that effect. A powerful principle of repulsion is manifestly introduced; and hence the igneous fluid, if it were separately exhibited, would assume a gaseous and expansive form. It must indeed possess astonishing elasticity.

Here is, therefore, the play of three forces;—the mutual repulsion of the particles of the igneous fluid—their attraction to the particles of the body with which they are combined—and the attraction of these to each other in consequence of their extension or displacement. The attraction of that subtle fluid is only the connecting link; its internal repulsion, and the internal attraction of its basis, are properly the antagonist forces. The effort of the fluid to dilate itself, increases with the measure of its accumulation; the tendency of the particles of the heated body to regain their quiescent station, augments with the degree of their distension. This equipoise must always obtain, and the latter force is consequently indicative of the former.

Thus are we brought by a close chain of induction to this important conclusion, that *heat is an elastic fluid, extremely subtle and active.* Is it a new and peculiar kind of fluid, or is it one with which, from its other effects, we are already in some manner acquainted? If any such can be discovered that will strictly quadrate with the phœnomena,

nomena, the spirit of true philosophy, which strives to reduce the number of ultimate principles, would certainly persuade us to embrace it. But in searching farther we may perhaps educe direct proofs of identity.

Heat and light are commonly associated. The materiality of light appears to be supported by irrefragable arguments, which I need not here repeat. It is emitted from luminous bodies with inconceivable force; but it must have previously been contained in them. Light has, therefore, two diſtinct modes of existence—that of projection—and that of combination. The former state only is that generally known and admitted. But all substances are capable of yielding light: collision—attrition—inflammation—the action of the electric shock—are the several ways by which the effect may be produced. Two lumps of quartz, struck or rubbed against each other, are made to discharge light: and the experiment succeeds not only in air, but under water and oil, and even within an exhausted receiver; an evidence that the light is derived from the interior mass of the stone. Other earthy bodies possess similar properties. though in different degrees. The sudden union of the oxygenous and hydrogenous gases, occasions a most copious flow of light. In passing to steam or water, those gases change their constitution, and set free the luminous particles which were latent in them. However variously obstructed, the oxygenous gas in every species of inflammation, sheds profusely its light. There is no substance but becomes luminous by the passage of the electric influence. This readily succeeds with wood, ivory, glass, wax, oil, water, air,* nay, the metals themselves. Nor, though we should admit the reality of an electric fluid, could such emission of light be with any justice referred to that source; for, in performing the circuit from the one side of the charged plate or jar to the other, it must evidently be restored, without alteration or diminution, exactly in the same state.

But the transit of light is the act of a moment. If its origin be concealed, so likewise is its termination.

* See Note XVII.

nation. When a body is opposed to a lucid beam, it reflects part, perhaps transmits part, and the rest seems lost. Yet is this portion not finally extinguished; it is only stopped in its flight, and united to the medium by attraction. The reflection and refraction of light have been investigated with accurate attention, and constitute the two capital divisions of the science of optics. But its *absorption*, a property still more extensive, has almost entirely been overlooked. To underſtand the matter clearly, we should examine the circumstances which mark the transition from reflection to refraction, and from refraction to absorption. The reflection and refraction of light, evidently testify corresponding repulsive and attractive powers, subsisting at certain distances between its particles and those of other bodies. Between the range of repulsion and that of attraction, there is some obscure limit, from which the change on either side is at first gradual. In the case of water it is rather nearer than in that of glass, and is the most distant in that of the metals. When the reflecting substance has an

even

even or polished surface, the dividing limit will form a parallel plane. The oblique motion of a ray of light impinging against a smooth surface, may be resolved into two motions, the one parallel, and the other perpendicular, to that surface. The latter only is subject to the superficial action; the vertical approach of a lucid particle is resisted, and at length stemmed; and the same repulsive energies being repeated, produce an equal and opposite celerity, which, compounded with the parallel motion, that has not been affected, occasions a reflex course, making an angle equal to that of incidence. Thus, the particle of light does not start back by a sudden abrupt resiliency, but describes at its flexure a small curve, convex towards the reflecting surface. After it has reached the apex of that curve, and is shaping its course in a parallel direction, there is a pause of hesitation, when the slightest disturbing force may be sufficient to destroy its ticklish poise; and by giving a bias either outwards or inwards, cause it indifferently to be reflected or refracted. The casual proximity, for example, of some

some minute depression in the surface will repel the ray, and the accident of an encroaching protuberance will attract it. Such irregularities must occur in every surface, since none is perfectly smooth, and the effect of polish is only to diminish to a certain degree the size of the natural indentings. The more oblique is the angle of incidence, the less must be the approach towards the surface. The perpendicular force of the ray being then smaller, is sooner extinguished. When it impinges more directly, it must, for a like reason, penetrate nearer. Between those extremes of approximation is situate the quiescent limit. If the flexure of the ray coincide with this, there is an equal chance, from the occurrence of a minute cavity or prominence in the surface, that it will be reflected or refracted. But if it either go beyond, or fall short of the boundary, the slight tendency in consequence to enter or to recede, will disturb that equality, and the scale will incline to the one side or the other. It is hence that the oblique rays are always most copiously reflected. In confirmation of this remark we may adduce a familiar fact. If I survey the image of

my

my face in water, it appears extremely faint; but if I view any thing reflected from the same surface at a very oblique angle, it will seem surprizingly bright. Trees and houses near the margin of a smooth lake, when seen from the opposite side, inverted in the water, look as distinct and vivid almost as the objects themselves. A variety of circumstances concur to show, that the metals exert, not only more distant, but more vigorous repulsions than glass or water; and hence, in the case of the former, the incident rays are arrested for the most part before they reach the limit of quiescence, and of course the chances of their being reflected are, by that extraneous influence, made to preponderate. If the surface of a body is entirely devoid of polish, its boundary of equipoise must, though in a less degree, be likewise uneven, and perhaps interrupted. In that case the prominences will have greater influence than the cavities, for the proximity will disclose and magnify their partial actions. The effect of distance is to soften, intermingle, and confound the disturbing forces. This remark admits of a simple illustration. If a bit of plane glass

glass has its polish destroyed, by means of emery or sand-paper, it will in general be impossible on the matted surface to trace the reflected images of objects. But if we hold it in a position extremely oblique with respect to the eye, the reflections will become perfectly distinct, and almost as vivid as if it had been highly polished. Those slanting rays glance at such a distance, that a sensible portion of the surface is brought at once into action, and consequently its successive irregularities, blending together their opposite influences, must nearly counteract and extinguish each other's effects. If the surface however be excessively rough, that balance of partial impressions cannot obtain; a small proportion of rays only will be reflected, and those few dispersed in every direction.

Thus we are able to form some idea of the nature of those delicate and abstruse operations which determine a particle of light to recoil, or to enter the substance of a body. But after, being urged by a general attraction, it has penetrated into the mass, its subsequent progress, through the ramified internal vacuities, is still liable

liable to interruption. If it chance to pass too near a corpuscle, it will be powerfully solicited by a partial action, and turned aside from its course; or if it encroach within a certain limit, its motion will be extinguished, and it will remain in a state of union. Such appears to be the cause of the *absorption of light*. Bodies are astonishingly various in their structure. The arrangement of the elementary points must be incomparably simpler in transparent substances than in opaque. The former may be loosely compared to an artificial plantation, the latter to a natural wood: in the one, an arrow shot in any horizontal direction has some chance to escape; in the other, its flight must be soon stopped. But between opaque and transparent substances there is no absolute distinction: none is strictly pellucid, and none is absolutely impervious to light. Gold-leaf, or a thin plate of ivory, is tolerably diaphanous, and a thick block of glass may be considered as opaque. The colour of the bottom of the sea becomes altogether imperceptible beyond the depth of 80 fathoms. With respect to transparent bodies the absorption of the rays of light is

greatest

greatest at the moment of their entering or emerging; for their motion being then feebler and less decided, is more easily affected by every impediment.

But the light which is absorbed into the substance of a body, does not entirely lose its innate activity; it continues to exert, among its own particles, a strong mutual repulsion. This is proved from various considerations. Light must evidently be discharged from luminous matter by some effort of a repellent power; but it would be contradictory to suppose such a power to proceed from the mass itself; for, in that case, how could the previous union have ever obtained? Besides, the particles would not radiate in all directions, but flow in lines perpendicular to the surface, especially if it was polished. A similar effect would take place if each projected particle of light was urged by the combined action of its adjacent particles. To explain the divergency of the rays, it seems necessary to admit likewise a lateral repulsion, which might spread them in all directions: that is, the particles of light must not only repel each other when lodged within a body,

body, but even after they have escaped and are actually in motion. In confirmation of this proposition, I might cite some curious facts respecting the inflection of light: it would lead however to much intricate discussion, altogether foreign to our purpose.

It thus appears, that light, while in the state of combination, possesses the distinguishing characters which must belong to the igneous fluid—extreme subtlety, powerful elasticity or repulsion among its own particles, and eminent attraction to those of all other substances. A coincidence, so striking in every point, might alone incline us to consider light and heat as identical. But such evidence, however seducing, is only presumptive; and fortunately the proposition can be supported by direct and unexceptionable proofs. I need mention only a single fact, which, duly weighed, will appear entirely conclusive. *If a body be exposed to the sun's rays, it will in every possible case be found to indicate a measure of heat exactly proportioned to the quantity of light which it has absorbed.* This statement is agreeable to common observation. A thin transparent substance, held in the sun-beams, scarcely

scarcely acquires any sensible heat; and the impression of the solar rays on the bright polished surface of a metallic body is equally feeble. A mercurial thermometer, and one whose bulb is filled with deep tinged alcohol, are very differently affected in the sun. The heat which dark-coloured substances conceive from the afflux of light, is well known. But, on closer examination, the principle above stated will appear to apply with perfect accuracy. The most delicate trials evince, that, in like circumstances, the elevation of temperature always corresponds with the greatest nicety to the degree of absorption. The experiment may be performed, after a variety of ways, by light enfeebled either in transmission or reflection. For instance, if the interposing of a plate of glass diminishes the acquired temperature of a body exposed to the sun by one-tenth part, the addition of another similar plate will occasion a farther reduction of one-tenth of the remainder; and so on, forming, as we should expect, a descending geometrical progression. The calorific action of the sun is observed to decrease, in proportion as he declines from the zenith; but this is not owing merely to the oblique incidence of his rays against the recipient substance; for, in the case of a globe, it must be unalterably the same. The light suffers a greater degree of diminution, according to the increased length of its passage through the atmosphere; and its different impressions will be found to agree precisely with the results derived from calculation. It were easy to multiply arguments and illustrations. But enough has, I presume, been stated to establish the conclusion, *that heat is only light in the state of combination.*

This theory, I will confess, is yet liable to some objections; but they are not formidable, and they seem to admit of a satisfactory answer. They chiefly refer to certain delicate chemical phænomena, which are produced by the single operation of light. They are indeed reducible to one principal fact—the extrication of oxygenous gas,—which takes place in the growth of plants, and in the partial revival of a few metallic oxyds and their solutions. But I would observe, that in strictness this property does not exclusively belong to light: the simple application of heat is capable, more or less, of producing analogous effects.

The

The peculiar energy of light may with reason be ascribed to its force of impulsion. Moving with inconceivable rapidity, the progress of its particles cannot be stopped without occasioning in the obstacle a vehement, though diffuse, reaction. Mechanical pressure or impulse, will appear to exert a very considerable influence in modifying or deciding the play of chemical affinities: and when these are nicely balanced, the smallest disturbing force may be sufficient to cause a new combination. Thus, in certain stages of their oxydation, the slight blow of a hammer will revive silver and mercury, with violent explosion. And why should not the stroke of light, in its gradual accession, silently operate in some degree a similar effect on the nitrate of silver? It is well known, that ivory, and many other absorbent substances, moistened with a dilute solution of silver in nitric acid, and placed in the sun, acquire a deep and permanent black stain; which can be imputed only to a partial revivification of that metal.

With regard to the influence of light in vegetation, it probably acts merely as a stimulant. To separate the oxygenous gas from the atmosphere, or from its nutriment of water, is a process perhaps essential to vegetable life; and the appulse of the rays of light must excite and invigorate all its functions. It is by many supposed, that light constitutes the green fecula of plants; but the arguments brought in support of this opinion, do not appear to me well grounded. Light seems requisite to the health of plants. Deprived of its beneficial energy, they become flaccid, and pale, and sickly. Their whiteness is only a symptom of disease. It may be produced by other causes, which can only introduce morbid affections. For example, the stalks of culinary vegetables are blanched by heaping them with earth; since in the effort to convert its trunk into a root, the plant suffers languor and topical debility.

It may be still objected, that the fluid of heat never displays itself in a separate collected state. This proceeds from its universal attraction to other matter. When once disengaged, it will stop its rapid flight only to enter into close combination with the obstructing body. We may discern some remote analogy in the case of certain fugitive gases, which cannot be confined but

by

by quicksilver, and which, the moment they come in contact with water, or other liquids, are absorbed and disappear.

It seems a constant law of gaseous fluids, that the mutual repulsion or elasticity of the particles, is inversely as their proximate distance from each other. Hence the expansive power which they display is proportional to the density; for it is evidently compounded of the forces exerted by the particles singly, that is, it must be in the joint ratio of the number of particles in any section, and the intensity of their repulsion. This principle, we may fairly presume, extends likewise to heat, or quiescent light. In the same body, the igneous fluid will, by the force of its elasticity alone, maintain an equal diffusion. In different communicating bodies, the quantities of heat dispersed among them, must depend on their respective attractions. If the attractions were proportional to the densities of the absorbent substances, the fluid of heat, which, by its internal repulsions or expansive action, balances those forces, would be distributed in the same ratio; that is, the quantity of heat would always be as the quantity of matter. But, if every substance exercised the same attraction upon heat, it would in all of them have an equal degree of condensation, or the quantity contained in a body would be proportional merely to the space it occupied. The measure of the heat which is lodged in different substances appears, for the most part, to follow some intermediate relation between the *weight* and the *bulk*. Its attraction therefore to bodies augments generally with their density, though not in so high a ratio. The density however of heat must in each case be determined by the *specific attraction* of the absorbent substance.* This specific attraction no doubt results from the collective energies of the primordial corpuscles, as modified by their peculiar internal arrangement; but, like chemical affinities, in consequence of our entire ignorance of the nature and effects of elementary structure, it can be ascertained only by actual experiment. Where a body suffers compression or condensation without altering its physical qualities, we may form indeed a vague estimate of the change introduced in its disposition

* See Note XVIII.

towards

towards heat. It is plain that the igneous fluid must extend somewhat beyond the real boundaries of the substance which contains it; and it seems probable that there is always a certain constant limit to this protrusion. We may suppose also, that the attraction is chiefly, if not wholly, exerted by the proximate corpuscles. As its distance therefore remains the same, a particle of heat will be held by a force proportional to the number of those particles that occur in any section; that is, as the square of the cube root of the density. Thus, if a body were concentrated eight times, its superficial concentration would be increased four times; but the mutual repulsion of the particles of light would now be eight times greater than before, and consequently its specific attraction would be reduced to one-half.

Hence the reason why a body evolves heat in being condensed; for its specific attraction is thereby diminished, and consequently, while in equilibrium with the contiguous bodies, it is no longer capable of retaining an equal portion of heat. Till that equilibrium be attained, the igneous matter which it previously held, must indicate a higher pitch or temperature. A very sensible warmth attends the compression of air, and the hammering of iron, lead, and other metals. In the latter, there is really induced a certain change of constitution.

When any substance changes its constitution or internal arrangement, it likewise changes its attraction for heat; and each successive transition is marked by a corresponding increase or diminution of that measure. It is generally less in solid than in liquefied bodies, and still less in these than in their vaporific expansions. Water furnishes an obvious illustration. In the form of ice its attraction for heat is one-tenth part less than before; in the state of vapour, that power is two-thirds greater. Hence the aqueous substance, exposed to the operation of the same heating cause, will not manifest a regular increase of temperature; the progression will be suspended at the several stages, by certain stationary intervals. Cold ice grows uniformly warmer till it has reached the point of congelation; a pause then ensues, during its conversion into water. But this being once atchieved, the series is now resumed, and the temperature of the

the water rises with equable ascent to the limit of boiling, where a stationary interval of still longer duration again occurs. And after the steam is thus formed, it is thenceforth susceptible perhaps of a boundless increase of temperature. At each successive station, therefore, a farther absorption of heat is required, to preserve the same temperature, or maintain an equilibrium with the contiguous matter. Nor does the efficacy of the heat applied, continue the same through the several interrupted spaces in the scale of temperature. It has greater influence on ice than on water, and still greater on water than on steam. The elevation of temperature which a substance receives from an equal accession, is in every case, except where a change of constitution takes place, reciprocally as its specific attraction. Thus, a portion of heat sufficient to raise the temperature of water 9 degrees, will produce on the same matter in the form of ice, the effect of 10 degrees; but on steam, it would occasion only a rise of 5½ degrees.

## CHAPTER X.

ONE of the most curious problems that has been attempted in chemical philosophy, is to discover the distance of the absolute zero, or the beginning of the scale of heat. This is effected by determining, in a variety of instances, the relation which subsists between the change of temperature and the corresponding alteration of specific attraction. The results, however, differ considerably; nor is it a matter of surprize, when we reflect on the extreme nicety of the question, and the uncertain nature of some of the data. We may reckon the mean determination at 750 degrees centigrade;* and this distance to the commencement of the absolute scale of heat, is exact enough for every speculative purpose.

If a body were to expand the 750th part of its bulk for each degree of increase of temperature, the elastic force of the heat combined with it

* See Note XIX.

would

would evidently continue invariably the same. Hence, on that supposition, the antagonist force, or the attraction of the particles of the body to those of heat, would likewise remain unaltered by their distension or enlarged separation. Such event, however, is barely possible; for we cannot conceive an attractive or repulsive power that is not some function of the mutual distance.

If the expansion of a body by heat be less than the 750th part of its bulk for each degree of the centigrade thermometer, the combined igneous fluid, being thus accumulated in a higher ratio than its dilatation, will, on the whole, be concentrated, and must therefore exert an increasing repulsive force. To counteract this, the attraction of the particles of the body to those of heat, must augment also in a similar manner, as they mutually recede, or as they are drawn aside from the quiescent limit. Such appears to be the constitution of all fixed substances, whether solid or liquid. On the other hand, when the expansion of a body for each degree exceeds the 750th part of its volume, the combined heat is progressively more dilated than accumulated, and will consequently

quently have its elastic force continually enfeebled. Its adhesion therefore to the corpuscles must also regularly decrease. And such is the constitution of the gaseous substances. The attraction of the integrant particles of bodies seems, at a certain distance, to reach its maximum, beyond which limit it again declines.

If the attraction of the corpuscles to the matter of heat increases or decreases uniformly, the corresponding expansions will be likewise equable. In solid substances, as the metals or glass, this appears to be nearly true; but in liquids, there is a very sensible deviation from the law of uniformity. As the attractive force, which balances the elasticity of the igneous fluid, tends towards its maximum, the successive augmentations that it receives become gradually smaller and smaller. Hence the expansions produced by equal additions of heat, form in general a rising progression. This is observed even in mercury; it is very perceptible in alcohol; and in water the successive increments of volume may be reckoned proportional to the distance from the point of congelation, and consequently the whole expansions from that

that point constitute very nearly a series of squares. In its transition from the liquid to a gaseous state, a substance must pass the attractive limit. When there is a great interval between the solid and the vaporous form, the expansions by heat are, for that reason, more equable. Thus, between the freezing and boiling points in mercury, the distance is 360 degrees; and in alcohol it probably exceeds 150 degrees, while the intervening space in water is only 100 degrees.

If the principles which we have stated be correct, they will enable us to penetrate some of the abstruse operations of nature. In the first place then, it is possible to determine the absolute elasticity of the igneous fluid combined with bodies. Take air for an example:—If you communicate to it one degree of heat, or the 750th part of the whole heat which it contains, it will expand the 250th part of its bulk; but if you now subject it to the additional pressure of the 250th of an atmosphere, it will shrink again into its former volume. Therefore those two forces must balance each other; or the 750th part of elasticity of the igneous fluid is equivalent to the 250th part of the

the elasticity of common air. Consequently the expansive force of the former is three times that of the latter. The same proportion seems to hold with respect to the other gases. The hydrogenous gas, which is the most distinguished by its properties, suffers an equal dilatation by heat as atmospheric air. But the igneous fluid contained in it is likewise of equal density, therefore of equal elasticity with that combined with air; for, if the hydrogenous gas has its specific attraction for heat ten times greater than that of common air, it is also ten times rarer.*

In the next place, we may calculate the density of the igneous fluid, or the quantity of matter which it actually contains. But to perceive clearly the grounds of procedure will require some attention. It has been already shewn, that heat, in the state of emission, constitutes light; and the laws of optics require that light, from whatever source it originates, must always flow with the same velocity. The refraction which a ray of light suffers on entering a diaphanous substance in an oblique direction, depends on the joint considera-

* See Note XX.

tion of its previous celerity and the intensity of attraction which it experiences. If in any case its appulse was more rapid, the deflection from its course, being effected in a shorter time, would be proportionally small. On such a supposition, the focus of a convex lens would retire to an unusual distance. But this is contrary to observation; and the eye, which is only a compound lens, is evidently fitted for every species of light. All the rays, therefore, must issue from their luminous sources with the same identical celerity; whether they dart from the sun, a candle, or a fire; whether they are elicited by the collision or attrition of hard bodies, or are discharged from a wide range of substances by electrical agency. It hence appears, that light must derive its projectile impulse from the sole operation of its peculiar elasticity while in the state of heat. Its motion is exactly similar to that with which an expansive fluid will rush into a vacuum. The velocity is not at all affected by the degree of previous condensation, but depends on the distending force compared with the quantity of matter on which it acts. It is the same velocity as what would

would be produced by the pressure of an homogeneous column, whose weight is equivalent to the measure of its elasticity, and therefore the same as that which would be acquired by falling through this height. If a fluid has its elasticity diminished by rarefaction, the mutual distance of its particles, or the space of action, is proportionally increased, and consequently the final effect, or the velocity generated, must continue the same. In different fluids, the square of that velocity is directly as their elasticity, and inversely as their density. Thus, hydrogenous gas would rush into a vacuum more than three times faster than common air, because with the same elasticity it is at least ten times rarer.* The vast celerity of light must be ascribed to its extreme tenuity, and prodigious expansive power. We are forced to suppose, that when bodies discharge it, they are thrown into a sort of convulsive state, having their adhesive attraction to it affected by momentary intervals of suspension, during which fits the luminous particles, being set free, are projected by their own intrinsic repulsions. Without ad-

* See Note XXI.

mitting

mitting this hypothesis, it seems impossible to explain the equality of motion which belongs to every species of light. However variously combined with different bodies as constituting heat, it is emitted from them all with the same rapidity. And such, we have seen, is the remarkable property of an expansive fluid when liberated.

Light travels from the sun to the earth in eight minutes, or at the rate of 200,000 miles each second. Its velocity, compared to that with which air rushes into a vacuum, is therefore = $\frac{200,000 \times 5280}{1350}$, or 782,000 times greater,* and the square of this, or in round numbers, 600,000,000,000, will denote the astonishing relative elasticity of light. But light or heat has been shown to be three times more elastic than the air with which it is incorporated; it consequently must have only the 200,000,000,000th part of the quantity of matter which that fluid contains: And if the usual estimate be just, it exists 500† times more condensed in water, whose combined heat must hence form only the *four*

* See Note XXII. † See Note XXIII.

*hundred millionth* part of its total weight. No wonder then, that all the various attempts to determine the ponderability of light or heat have hitherto proved fruitless.

But we may still venture a step farther, and ascertain the secular or annual expense of the solar substance, occasioned by the copious and incessant emission of luminous matter. A blackened hollow ball, of any dimensions, and filled with any sort of liquid, if exposed in calm air to the undiminished force of a vertical sun, would acquire a heat of about ten degrees. Such a globe, one foot in diameter, and containing water, would be found, after that action is removed, to lose by cooling, for every four minutes, the 100th part of its whole excess of temperature. This is, therefore, the measure of the calorific power of the sun; it communicates to a globe of water of a foot diameter, at the rate of one-tenth of a degree of heat in four minutes. In different spheres, the accession of heat is evidently as their surface only, and consequently its effect on the whole mass will be inversely as their diameter. If the diameter of the globe holding water, were extended to a mile, the

the impression which it would receive from the incident beams, would only amount to one degree in 3520 hours, or 5280 × 40 minutes. In that space of time, therefore, a portion of heat is received equal to the 300,000,000,000th part of the whole aqueous matter; for 750 × 400,000,000 = 300,000,000,000. And, since 3520 is to 5280 as 2 to 3, or in the ratio of a sphere to its circumscribing cylinder; if the heat absorbed each hour were reduced to the density of water, it would form a film of the 300,000,000,000th part of a foot in thickness. In one year, this would accumulate to the 34,223,000th part of a foot. Such must be the thickness of luminous matter that would fall on the whole concavity of the earth's orbit, if, to assist the fancy, we borrow the ancient notion of crystalline spheres. If that expanded igneous coat were conglomerated together, and still of the density of water, it would be found, by a simple computation, to form a globe of 49.878, or very nearly fifty miles in diameter. Hence the light which the earth receives annually from the sun, is equal in weight to a sphere of water 139 feet in diameter. We may estimate the density of the solar substance at 3½ times less than that of water: wherefore, by the continual discharge of light, the sun will suffer a waste from his surface of the depth of one foot only in the space of 700 years:—a quantity surely too inconsiderable, compared with his vast mass, to occasion any sensible relaxation in the planetary motions during the countless revolutions of ages.

Another consequence, equally striking, which we derive from those principles is, that our earth must grow continually warmer by the accession of the solar rays. Whether those rays reach the surface of the ground, or lose themselves in the clouds, their influence will ultimately be the same. They must soon come to unite with the general mass of the globe; for, beyond the boundaries of our atmosphere, there are no gaseous fluids to disperse the circulating heat indefinitely into space. We have even data for ascertaining, at least within the limits of probability, the very measure of effect produced by this absorption of igneous matter. It was already observed, that a ball of water perfectly insulated, and of a mile in diameter, would acquire one degree of heat from the full impression

impression of the sun in the lapse of 3520 hours. Consequently, a ball likewise of water, but 7985 miles in diameter, would take 3206 years for a similar effect. If our globe, therefore, consisted entirely of water, it would grow one degree warmer in the period of 3206 years. But it is four or five times denser than water,* and is of course denser than any of the known primitive earths. It seems more akin to the metallic oxyds, and the phænomena of the magnetic needle afford a strong presumption, that the internal body of our globle is ferruginous. The density of the rust of iron is about 4½, and its specific attraction to heat one-sixth. Hence the igneous fluid which it contains is only three-fourths of the density of that combined with water. The mass of the earth will, therefore, after that proportion, be sooner heated than if it consisted of water; or its medium temperature will mount at the rate of one degree in 2405 years.† Such a conclusion is entirely consistent with the testimony of past ages. The climate of the middle and northern parts of Europe has become gradually

* See Note XXIV. † See Note XXV.

milder. Nor can this be referred to the effects of human industry, in clearing the surface and improving the soil. Those beneficial labours have some tendency, indeed, to diminish the inequality of the seasons; but they can have no influence whatever in altering the average of temperature.

Light is thrown from luminous bodies nearly in the same manner as water is spouted from an aperture in the side of a vessel. The direction of the flow is that which results from the conjoint pressures by which the particles are urged. When not affected by extraneous causes, such as the depth of the discharging orifice, it is always perpendicular to the bounding surface. Thus, if a bladder filled with water be punctured with a needle, and then squeezed, a jet will be formed exactly at right angles to the small space around the hole from which it issues. Reasoning from analogy therefore, we should conclude, that light ought likewise to be projected in rays perpendicular to the surface. But this is contradicted by observation; for a square bar of iron made red-hot, will shed its rays in every direction. In the case of water, the motion is begun at some small depth below

below the surface, and the particles affected are impelled by an action that extends to a sensible distance on either side. The direction of the jet is therefore not determined by the individual position of the lips of the orifice, but by the general contour of a certain encircling space. The operations of the igneous fluid are probably more concentrated. The particles of light may, consequently, be projected in lines perpendicular to each minute portion of surface. But it is well known that every surface, even the smoothest and most uniform, when strictly examined, appears full of irregularities: And hence on this principle, merely, the rays of light may be dispersed in every direction. The supposition however seems very forced. High polish is found to diminish the size of those inequalities to such a degree, that, in the phænomena of optics, they mingle and equalize their effects, losing almost entirely their deranging influence. We might therefore expect, that the same process should destroy the radiating property of luminous bodies. Besides, it cannot be doubted that inorganic substances, nicely considered, have their internal structure perfectly regular. Their fracture, or the abrasion of their surface, only discovers a range of crystals, of different sizes, and more or less compounded. But it is obvious, that the crystalline facets must have the same positions, however variously grouped. Nor can those several positions be numerous; for the primitive crystals have small variety of angles. Consequently, if the rays of light be thrown in lines perpendicular to the facets, they will not spread on all sides, but will affect certain particular directions.

We have therefore to seek some other mode of explaining the dispersive radiation of light. Nor is it difficult to discern what appears to be the true cause of that phænomenon. When the heat united to a body is for a moment set loose, the particles at the surface, being actuated by a general repulsion, are at the same time impelled outwards and urged by a lateral force. The particles thus shot off from any point, will not proceed in a concentrated stream, but spread out in diverging lines; and, as the pressure is equal on every side, the directions of their flight must likewise make equal angles with each other. The uniform radiation

radiation of light, independently of the nature of the surface, is therefore a necessary consequence of the liberation and developement of its elastic powers.

But though it is a received opinion, that light radiates from luminous bodies with equal dispersion, this proposition will appear on examination very far from being accurate. If a shining flat surface placed at a considerable distance, be turned more and more obliquely to the eye, its brightness will continue nearly the same: were the rays however equally copious in every direction, it is evident that the degree of illumination ought to grow more and more intense, in the successive positions of the surface; since the eye receives still the same quantity of light, while the optical magnitude, by reason of the increasing obliquity, is always contracting. The brightness of a luminous surface would be in the inverse ratio of the cosine of its inclination, or as the secant of that angle. Hence, a red-hot ball should appear the darkest about the centre, and extremely bright near the edges. But this is quite contrary to fact, for at a remote distance the ball is not distinguish-

able

able from a flat luminous disc. It hence follows, that light is emitted less copiously in the oblique directions, and that the density of the rays is nearly as the cosine of their deviation from the perpendicular. The cause which I would assign will perhaps seem too refined; yet it is agreeable to analogy, and entirely consistent with the phænomena. The particles of light are projected at first with equal radiation, but they are not suffered to pursue their original course. After the pulse during which they acquired their motion, has terminated, they become subjected again to the attraction of the body, and therefore they are bent back from the perpendicular direction, exactly in the same manner as oblique rays passing from a denser to a rarer medium, are refracted. And it is easy to prove, that the density of each pencil of light will be as the cosine of the emergent, divided by the cosine of the refracted or final, angle.* If the attractive force be considerable in comparison with that of diaphanous substances, the emergent angle will in every case be small, and consequently its cosine will not sensibly differ from the radius.

* See Note XXVI.

Wherefore,

Wherefore, the density of the light emitted would be very nearly as the cosine of its inclination. Thus, if the attraction was denoted by 2, which is not much different from that of glass, the extreme angle of emergence would only be 30°, and its cosine = .866. On this supposition the outer rim of a red-hot ball would only be about one-seventh or one-eighth fainter than the centre; a difference too small in general to be distinctly noticed.

CHAP.

## CHAPTER XI.

THE entire correspondence between theory and observation affords the most convincing evidence of the justness of our principles. It is therefore the same subtle matter, that, according to its different modes of existence, constitutes either heat or light. Projected with rapid celerity, it forms light: in the state of combination with bodies, it acts as heat. Under this latter modification, it is more immediately the object of the present inquiry.

The igneous fluid absorbed into a solid substance, is not immoveably fixed and incapable of circulation. Disturbed by any external cause, it again diffuses itself, and restores the equilibrium. The particles of heat contained within a body, being attracted equally on every side, are left freely to exert their own expansive powers. If accumulated in one part, the increased elasticity there will occasion a flow towards the other parts.

But

But though, in the circulation of heat, the substance which contains it is absolutely passive, the internal motions of that fluid must experience prodigious impediment and detention. Without such obstruction, its diffusion would be to sense instantaneous, having almost the celerity of light itself. Had this been the constitution of nature, it might amuse the fancy to contemplate for a moment its vast and tremendous consequences. An uniform and unvarying temperature would have pervaded the globe: no distinction of climate, no vicissitude of seasons, and no grateful alternation of day and night. The azure vault of heaven, perpetually serene and cloudless, would lose its animated charms. If snow and hail would be unknown, so likewise would the refreshing influence of rains and dews. The face of the earth would present one monotonous picture of sterility: no verdure to relieve the eye, no vegetation, and no sustenance for animals. All the springs of life would be locked up. The beneficial effects, the very existence, of artificial heat, would for ever have been concealed; for, the instant it was generated, it would spread and ingulph itself in the general mass.

The resistance that heat suffers in circulating through the interior of bodies, indicates a prodigious expenditure of force, which must be consumed in causing a multiplicity of irregular collateral motions. This resistance might proceed either from derangements among its own particles, or among those of the containing body. But the first supposition is in the highest degree improbable, since the expansive energies which heat displays, are incomparably superior to its gravity or quantity of matter. If, working its devious traverse or ramifying progress through a solid substance, it even spent in the various windings and doublings, in successive accelerated or retarded movements, one million times the force sufficient to produce a direct continuous flow; the transfusion of heat would still be many million times slower than what is actually observed. But probability, however strong, is always unsatisfactory; and the same conclusion is established by an argument quite incontrovertible. It is well known, that the resistance which a fluid encounters is proportional to the square of its velocity. Consequently, if heat owed the delay and impediment

diment which it meets with in permeating bodies, to an involved series of internal motions that alternately grow and expire again among its particles, the rate of its communication or diffusion would be as the square root of the difference of temperature or the force expended in surmounting those obstacles. If the celerity of dispersion, for instance, was doubled, there would be double the number of irregular movements to be produced, and these likewise twice as rapid; wherefore the aggregate momentum would be four times greater, or there would be required in that ratio the expansive pressure resulting from the local accumulation of heat. It is however an ascertained principle, that heat is conducted through the same substance, exactly in the simple ratio of the excess of temperature. We hence see the impossibility, from any supposed system of multiplied internal motions among the particles of the igneous fluid, to account for the resistance which it experiences in its transfusion.

It follows then, that the resistance which heat encounters in its passage through the interior of bodies, originates wholly from certain reiterated subsultory

subsultory motions or expansions, impressed on the connected particles of the recipient. This explication will perfectly satisfy the conditions of the question. Since the quantity of matter affected continues in every case the same, the number and relative extent of the internal displacements, occasioned by the communication of heat, will likewise remain unaltered. But in all similar motions, the velocities are proportional to the actuating force; and, consequently the rate with which heat is conducted into the general mass, will be exactly as the excess of temperature. It is manifest also, that the final expansions among the corpuscles must be in the same ratio, or that of the degree of heat which is received. And, that the velocities are thus proportional to the spaces, is the character of vibratory or isochronous motion.* Nor can we suppose, that each portion of the recipient substance attains its ultimate arrangement by a single though large oscillation; for such would be incomparably too slow to explain the prodigious resistance encountered. We must, therefore, admit that, during the commu-

* See Note XXVII.

nication

nication of heat the particles of the body are agitated by a series of minute pulsations, or undergo successive partial expansions, with intervening pauses. But this curious inference deserves more particular discussion.

Let two masses of unequal temperature, and situate at A and G (fig. 13), be made to communicate by means of a string of corpuscles, or a narrow cylinder of solid matter. If the mass at A be the hotter, its excess of heat will be continually transferred along the slender connecting rod from A to G; and, if the absorbent mass which touches at G be supposed to be incomparably larger, the heat thus deposited, being extremely dilated, will produce no sensible impression. It is obvious, that the extremities of the rod must have the same temperatures as their respective contiguous masses, and consequently that the temperature at G may, without sensible error, be considered as permanent. From A to G, there must be a gradual transition of temperature. Conceive the rod to be distinguished into a number of equal portions, AB, BC, CD, &c. and let the temperature at A be denoted by the perpendicular AH; if the progressive decline of temperature along the chain of corpuscles be strictly uniform, the temperatures at B, C, D, &c. will be expressed by the ordinates BI, CK, DL, &c. bounded by the straight line HG. But, in transferring the heat from A to G, each portion of the connecting rod must constantly both receive some heat and deliver it. While it acquires heat, the conducting substance will expand; and while it parts with heat, this will contract. But it seems impossible to conceive the opposite effects of dilatation and contraction co-existing in the same portion of matter. Heat must, therefore, be received and again deposited, by two distinct though consecutive acts. While the portion AB, for example, has the temperature and corresponding expansion AH; at that same instant, the portion BC has the temperature and corresponding expansion CK. In the next instant, acting on each other, they produce an equilibrium, or mean temperature, BI: AB loses the heat OI, and BC gains the heat KQ; the former suffering a partial contraction, the latter receiving a partial dilatation. In the third instant, the portion BC, with its temperature thus augmented to CQ, comes

comes to sympathize with CD, which has its temperature just reduced to DL. The two rectangles BQ, CL, melt into the single rectangle BS, and thus raise the temperature of CD to DS. Pursuing the same mode of reasoning, we trace the successive transfer of heat along the whole line of corpuscles. Each of the composite units of the chain must alternately undergo a hot and a cold fit. Thus, the temperatures of the portions AB, BC, CD, &c. must vibrate between AH and AZ, BI and BP, CK and CR, &c.; and, during each pulsation, the differences OI, KQ, LS, are respectively transmitted, by one remove, to the next adjacent stations. In the same conducting substance, the duration of the pulses will be proportional to the spaces affected. The impression will travel along the chain with the same celerity, as that with which motion would be conveyed. After the undulatory tide of heat has once arrived at G, a portion expressed by the differential rectangle ZHOI, will be delivered at each pulsation. Consequently, in a given time, the absolute quantity of heat communicated to the mass at G, will be simply as ZH, the altitude of that rectangle, or the variation of the successive ordinates, AB, BC, CD, &c.

From this view of the subject, the several circumstances on which depends the conducting power of bodies, may with great facility be deduced. I shall enumerate five capital points.—1. Other things being the same, the measure of heat transmitted is proportional to the excess of temperature: for OI is evidently as AH. This conclusion agrees perfectly with observation.—2. The communication of heat is inversely as the length of the conducting rod: for, while AH remains the same, OI is inversely as AG. This inference also corresponds with experiment, though the fact is not so easily or satisfactorily brought out.—3. The rate of transmission is compounded of the density of the conducting substance, and its specific attraction for heat; or it is proportional to the quantity of heat contained in a given space. For, at each pulsation, the final space FG communicates a charge of heat corresponding to the temperature YG or OI.—4. The celerity with which heat shoots along, is inversely as the square-root of the altitude of a column of the same density,

sity, and whose pressure is equal to the elasticity of the conducting substance. This appears from the principles of dynamics, since the propagation is made by vibratory impulsion.*—5. The flow of heat is proportioned to the breadth of the primary spaces, into which the conductor naturally divides itself. For, at every pulsation, a quantity of heat expressed by the differential rectangle HZIO, is transferred. But the time of a pulsation is as its extent AB; and therefore the rate of discharge is in the proportion of OI or AB. Thus, the division of the conducting rod AG into those vibratory portions, is not an arbitrary conception, but founded on its peculiar constitution. Every substance, when accurately examined, appears to consist of elementary crystals or fibres; and were these repeatedly decomposed, we should, no doubt, arrive ultimately at minute plates or corpuscles of a certain determinate thickness.

From the joint consideration of those five circumstances is derived the measure of the communication of heat. Their combined effect may be conveniently expressed by help of an algebraic

* See Note XXVIII.

formula. Let $h$ denote the excess of temperature in the mass at A, $l$ the length of the conductor, $m$ its density, $a$ its specific attraction for heat, $s$ the breadth of its primary intervals, and $e$ the height of the column corresponding to its elasticity. Then the variation of temperature at each pulsation must be $= \frac{hs}{l}$, and consequently the accession of heat $= \frac{hs}{l} \times ams = \frac{hams^2}{l}$; but the time of a pulsation is $= \frac{s}{\sqrt{e}}$, and therefore the quantity of heat delivered at G in a given time, is equal to $\frac{hams^2}{l}$ divided by $\frac{s}{\sqrt{e}}$, or $hams \times \frac{\sqrt{e}}{l}$.

In different substances, $am$, or the density of heat, varies comparatively little; and from what I have ascertained in a few instances, I am disposed to think, that the value of $e$, or the equiponderant column, in the whole range of natural bodies, is confined within moderate limits. The expression $s \times \frac{h}{l}$ therefore will, in general, be a near approximation to the rate with which heat is transmitted. The point on which chiefly hinges the

the conducting power of any substance, is the breadth, *s*, of its primary intervals. The nature of those intervals, it is perhaps beyond our penetration to discover, though their reality seems unquestionable. They may depend on the peculiar relation of the body to heat: they may be the vacuities, or spaces, which divide the ultimate atoms. On the latter supposition, since the metals convey heat remarkably faster than glass, they must have their primary intervals much wider, and consequently their elemental corpuscles proportionally larger. That metals act at greater distances than glass, is indicated by a variety of phænomena.

When, without the intervention of any connecting rod, a hot body is made to touch another body of the ordinary temperature, the excess of heat diffuses itself in a descending progression, the rate of communication diminishing, as the space affected extends successively with larger gradations. Thus, if the same heat be constantly applied at the one extremity A (fig. 14), of the cylindrical substance AB, the effect will first penetrate to E, and the temperatures of the contiguous strata will be denoted by the indented line CE. When the undulating current of heat has reached F, the corresponding scale of temperatures, CF, has evidently a gentler slope, and smaller indentings than the preceding. After the communication has arrived at the extremity B, the subsequent efforts will be spent merely in raising the temperature of the remoter portions. The scale of gradation will mount from B to the position CG, and its flexures will continually soften away as it approaches to the situation and character of the parallel straight line CD, which is its ultimate limit.

Where the heat is transferred by help of an intervening substance, the communication soon becomes equable and constant. But when it is propagated directly, the mutual relations, in the scale of temperature, are continually altering. The pulses now perform a double office: heat is not only conveyed to the remote parts of the absorbent mass, but portions of it are necessarily distributed over the intermediate space, in proportion to the distance from its source. Those partial deposits of heat are indeed necessary,

necessary, to maintain the regular extended series of temperatures.

Such then is the recondite process, which theory unfolds, of the communication of heat. In the case of solid substances, the general principles will not require any modification. But when the conducting medium is fluid, the mobility of the affected particles will so derange the mode of operation, as almost entirely to change its nature: for, the proximate portion of fluid, dilating in proportion as it receives heat, is gently forced to recede; and being likewise specifically lighter, it rises to the surface. The heat thus quickly spreads through the buoyant mass in horizontal strata, the hottest particles occupying the highest place, and the rest arranging themselves according to their respective degrees of temperature. The subsequent internal diffusion of this heat is performed very slowly, and with extreme difficulty. The heat is made to descend according to the general principle, by the successive transfer of minute differences from stratum to stratum. By the continual ascension of the heated portions of the fluid, and consequently the incessant

cessant renewal of contact, the heat is abstracted from an immersed solid with remarkable rapidity; but its circulation afterwards through the fluid, is in most cases tedious and imperfect. The bottom of the fluid acquires in general but a very small part of the heat of its surface. If the solid be supposed colder than the fluid which it touches, similar effects will be produced, but in an inverted order. The secondary motions will be directed downwards; since the portion of fluid in the vicinity of contact, being cooled, will descend by its superior gravity. The transfusion of heat, or the subsequent act, is performed in the same manner as before. Nor will the case be materially altered, whether we imagine a solid body plunged in the fluid, or another fluid, or a portion of the same fluid, suddenly introduced.

If the mobility of their particles contributes so greatly to the propagation of heat in liquids, that property, in a much higher degree, must extremely facilitate such communication and diffusion through the gaseous fluids. These elastic media, from their tenuity and expansibility, are susceptible of the most rapid agitation. But every species

species of motion seems accompanied with the transfer of heat. The gases are not merely capable of receiving the same impressions as liquid substances; they are fitted eminently, by their peculiar constitution, to acquire most extensive internal oscillations. It would however be unsafe at present, to push our theoretical disquisitions any farther. We must again appeal to facts, and prosecute the inquiry by the light of experiment. We now proceed to analyse the process, by which an insulated body disperses its heat in the surrounding atmosphere.

CHAP.

## CHAPTER XII.

IT was shown, that a hot or cold surface propagates its influence with astonishing celerity through the air, only by exciting some peculiar energy in that active medium. There are two modes, however, by which we may conceive this rapid discharge to be effected. It is, indeed, essential to suppose an aërial motion diverging from the source of action; but such motion may consist, either in the continued flight of the same particles, or in the successive transfer of agitation, by a vibratory impulsion which shoots along a chain of particles or through the general mass of fluid. In the former case, the heat would be remotely dispersed by the actual migration of the particles affected; in the latter, it would be communicated, from particle to particle, by a series of minute oscillations. Let us now inquire which of these hypotheses—whether the idea of a pulsatory transfer of heat, or that of a continuous flow of of heated matter—is most agreeable to the phænomena, and consistent with the known laws of physics.

We feel some repugnance to admit the supposition of aërial currents projected from a hot body. Experience seems to prove the very reverse. The air constantly flows from the door of an apartment towards the fire, and there, becoming heated, it makes its escape by the chimney. This incessant motion has a visible effect on the flame of a candle. But, if hot air streamed back into the room, could it elude observation? When a feather or a woollen rag is thrown into the fire, it quickly diffuses a strong empyreumatic smell. Yet, of smells, air is the proper and only vehicle. From the contact of that fluid, the odorous substances must derive their volatility, and they are gradually dissolved and transported through the atmosphere. That some air flows back from the fire, cannot, therefore, be denied. Nor is it any ways incongruous, to suppose the existence of two opposite currents. The bending of the flame of a candle proves only an excess of force which tends towards the fire. But wc might presume, that the

the current which feeds the combustion is more powerful than the one in a contrary direction, since it likewise supplies the portion of air that rises intermingled with the smoke. Nor can it be urged as an insurmountable objection to this hypothesis, that the air of the room would be contaminated and rendered unfit for respiration, by the continual admixture of the refluent streams which had lost their oxygene and contracted carbone in consequence of touching inflamed matter. We are not obliged to admit, that the air thus projected from the fire had been in contact with it, or assisted in combustion; for the mere proximity of the live coal might be sufficient to generate those regressive motions.

If the phænomena of radiation were occasioned, however, by the actual flow of hot air, it would be necessary, we found, to ascribe a prodigious velocity to the current. We might imagine, therefore, a multitude of slender streamlets diverging in all directions. Such aërial filaments, from the rapidity of their flight, would perforate, without deranging, the flame of a candle; and those perforations might be so extremely minute

as

as to become absolutely invisible, or have no other effect than somewhat to dilute the brightness of the flame. Nor is the idea of streamlets of air darting along with undiminished force, wholly incompatible with the received doctrine of fluids. The resistance which a body experiences in moving through a fluid medium, abstracting from the figure of its anterior portion, depends merely on the measure of its transverse section, and is not in any degree modified or augmented by the extent of its parallel sides. Such at least are the obvious deductions of theory. But, we may imagine streamlets so extremely slender, that the resistance which they encounter shall almost vanish; and the lengthening filament of particle succeeding particle with accumulating impulse, will then suffer comparatively no sensible impediment in its course.

Yet a little reflection destroys this specious argument. The supposed narrow currents of air cannot with accuracy be compared to solid rods: the mobility of the fluid particles must evidently occasion endless erangements; and, even admitting that their mutual cohesion is sufficient to maintain

maintain a permanent continuity, the motion would incessantly deflect from its primary direction, and describe a tortuous or serpentine track. The anterior portion of the streamlet, suffering the chief impediment, would gradually fall back and agglomerate; and, in this form, meeting with additional obstruction, it would spread out and divide into new filaments. This statement is perfectly agreeable to observation. A stream of air thrown forcibly from the pipe of a pair of bellows, soon scatters itself and seems lost in the general mass of atmosphere. We need not seek to impute this dispersive effect to the condensation, which must have preceded the extrusion of the air, and impressed it with a certain mutual repulsion or divergency, at the moment of its escape; for the same phænomenon is remarked in water, which possesses compressibility in such a very inferior degree. A river that discharges itself into a spacious lake, quickly communicates and wastes its impulsion, and relaxes the swiftness of its flow.

But the hypothesis of projected streamlets is liable to another objection perhaps still more formidable.

formidable. The theory of fluids, so defective in many respects, is more particularly imperfect in what concerns resistance. A cylinder, moving through a fluid in the direction of its axis, suffers a constant retardation proportioned to its extent of surface. Nor does this proceed from irregularity of shape, or want of polish, but merely from a sort of modified friction. The portion of fluid with which the sides of the cylinder come successively in contact, is thrown likewise into motion, and therefore consumes gradually the impelling force. The expence thus sustained must depend on the velocity of the affected particles, and the extent of the cylindrical surface. And a similar resistance will obtain, whether the penetrating body be a solid, or only a column of fluid. The narrower is this column, the greater must be its surface, compared to its mass; and, if the celerity be likewise augmented, the obstruction which it encounters must, on both accounts, increase in a high ratio. The supposition of aërial streamlets projected from a surface in the act of cooling, is thus pressed on all sides by insurmountable difficulties. The impulsion of those slender filaments would, in grazing through the atmosphere, soon relax and expire; contrary to the conditions of the phænomena, which require a continued and uniform rapidity of flight.

But the question, viewed in another light, may be brought to the decision of experiment. If currents of air were actually projected from a surface in the act of cooling, they would exert against it a powerful re-action. This follows from the primary laws of motion; and analogous effects are observed in a variety of instances—in the recoil of fire-arms, the ascent of sky-rockets, and the revolution of luminous wheels. But, if the hot surface excited merely a vibratory impression in the atmosphere, no retrograde action could take place; for the portion of air contiguous to that surface, being alternately dilated and condensed, the opposite forces thus evolved must destroy each other. In the first instant, the surface would be pressed back; in the second, drawn forward; and so forth, in repeated succession. Those equal and contrary efforts ought, therefore, to cause a mutual extinction.

The elastic force developed in the inflammation of

of gunpowder, is employed merely in driving forwards the gaseous vapour. On the supposition of projected streamlets, however, the re-action has a double office to perform: for the flow of air from the hot surface necessarily implies the existence of currents returning in a contrary direction; and to maintain the general equilibrium of the atmosphere, the quantity of motion tending towards the source of action must, though differently composed, be equal to that which darts from it. This direct flow, and the correspondent diffuse reflux, are both of them produced by the exciting force, which must, therefore, exert a re-active pressure, under equal circumstances, twice as great as in the case of the simple discharge of vaporous matter.

But if the opposite surfaces of a plate uniformly heated were of the same nature, it would still be impossible to distinguish the re-active effect of projection; for the pressure on the one side would exactly counterbalance that on the other. If, however, a surface of glass or pigment be made to act against a metallic surface, its influence

 ought

ought greatly to preponderate. Hence we are enabled to appeal to observation.

## EXPERIMENT XLI.

Having drilled a small hole near the edge of a circular piece of planished tin of about three inches in diameter, and painted one side with a smooth coat of China ink, I suspended it vertically from the ceiling of a close room, by help of a fine silver wire. On approaching the flame of a candle to heat the plate, it maintained its position, without deflecting in any perceptible degree from the perpendicular.

If the painted surface had recoiled only the twentieth part of an inch, this quantity would have been distinctly visible. But the line of suspension being about eight feet in length, the reaction could not, in that case, exceed the two thousandth part of the weight of the small plate. We may, therefore, confidently presume, that no projective force had been actually exerted.

## EXPERIMENT XLI.

Having removed one of the scales from a delicate balance, I substituted the tin plate, suspended in a horizontal position by three silver wires, and poised it accurately by putting weights in the other scale. On holding immediately under it a red-hot poker, it mounted upwards, with a force equal to about two grains. The same precise effect took place, whether the metallic or the painted surface was uppermost. After the plate had again cooled, it regularly descended to its former station.

It is obvious, that the buoyant tendency remarked in this experiment has no relation to a supposed projectile force, since it is altogether independant of the nature of the surface. Nor can it be imputed to any waste which the pigment might sustain in consequence of being intensely heated; for, a piece of glass, covered on the one side with tin-foil, exhibits the same property. The loss of weight is evidently occasioned by the slow yet constant ascent of a prolonged column of air, which, acquiring heat from its vicinity to the plate, becomes dilated, and therefore specifically lighter. No difference was perceived, whether the metallic surface was the upper or the under one; but, if that difference had amounted only to the tenth part of a grain, it must have been visible. Had projected streamlets any real existence, their re-action would have been very considerable. We shall afterwards find that, if the whole air in contact with a surface of paper or pigment, were to flow, charged to the same temperature, the ordinary consumption of heat would require a velocity of a foot in about five minutes. Consequently, if those supposed streamlets had the celerity with which air rushes into a vacuum, or 1350 feet in a second; only the 405,000th part of the contiguous air must at any time be affected. Hence the recoil of a painted surface would be double that quantity, or the 202,500th part of the pressure of the atmosphere; and, if the other surface was metallic, the difference of recoil, or the observable effect, would be diminished by one-eighth, or would be equal to the 231,428th part of the same incumbent weight. Therefore, between the two positions of the heated plate,

plate, a variation of weight should be perceived equal to the 115,714th part of the pressure of the atmosphere, and which would, in the present instance, amount to near six grains.

I shall now close this laborious discussion. It clearly results, that the idea of rapidly projected streamlets, though specious in some respects, is incompatible with the laws of fluids, and directly refuted by experiment. We are therefore compelled to embrace the only alternative, and to refer the diffusion of heat through the atmosphere, to the vehicle of certain oscillations, or vibratory impressions, excited in that elastic and active medium. Our next object of research, is to discover the nature, the origin, and subsequent propagation of these aërial vibrations. The theory of waves, whether superficial or internal, is still very defective; but the general facts are well understood, and will serve, if judiciously weighed, to correct and expand our notions concerning the physical operation.*

Imagine a string of connected particles to be urged at one extremity in the direction of its

* See Note XXIX.

length, by a gentle and slowly accelerated pressure. The momentary increments of force will be communicated successively from the first particle to the second, and from the second to the third, thus extending along the entire line. By the time, therefore, that the whole effect is imprinted, each particle will have acquired its equal share of impulsion, and the aggregate chain, cohering throughout, will move forward with the full momentum. But if we suppose the first particle to receive a sudden stroke, the shock will be transferred along the string without diffusing itself by an uniform partition. Instead of acquiring a progressive motion, the line of particles will be thrown into a state of vibration. Of this we have a familiar example: for, having placed a number of ivory balls, of the same size, to touch each other in a right line, if I give the first one a smart blow, the last ball, feeling the pulsation, and suffering no re-action, will start forwards, while the rest will retain their respective positions.

Percussion, or the sudden application of force, must, therefore, excite a vibratory commotion. But this peculiar agitation will be heightened, if the substance affected is either of great extent, or

is

is a slow conductor of impulse. The elastic fluids are remarkably subject to such impressions; and in the boundless expanse of atmosphere, every circumstance concurs to augment its internal undulations. The spaces, stretching from the centre of action, suffer in succession alternate contractions and dilatations. The breadths of those spaces, and the intensity of their modifications, depend on the quality and the degree of the exciting force. If we consider a series of consecutive particles, it is not requisite to suppose, that while A (fig. 15), is made to approach B, in the same instant C recedes from B and approaches D; the proximate intervals being thus alternately condensed and expanded. During the time that the particle A makes its approach to B, B may tend towards C, C towards D, though with a power continually diminishing, till the neutral point H; after which, the successive particles, as far as P, will mutually recede. Thus, a dilated space will extend from H to Y, to which will succeed an equal space similarly condensed. It is obvious, that the distance of the neutral point from the origin of pulsation, or half the breadth of the internal wave,

wave, must depend on the time of excitement; for, in proportion as the primary action is slow, its impression will have penetrated farther through the fluid, before the accelerating force is expended. If, for example, we conceive the duration of shock to be tripled, the limit of vibration will advance from H to Y. The quantity of shock is the joint result of its duration, and its intensity or momentary increments. The same elements, and the same transferred impulses, compose the measure or momentum of each pulsation; and the degree of alternate contraction and dilatation is evidently proportioned to the intensity of the original stroke.

Both of these circumstances are finely illustrated by the phænomena of sound, which consists in certain vibratory commotions of the atmosphere. The celerity of all sounds is almost invariably the same; their peculiar qualities depend on their loudness and their tone; the former corresponding to the intensity of the aërial pulse, and the latter to its duration. Thus, a musical chord will produce the same note, with whatever force it vibrates; the only difference lies in the degree

degree of loudness or swell. But if, by increasing the tension, we quicken its oscillations, the appropriate note will become continually sharper. Every sound, though apparently simple, is actually composed of regular tones; but these tones are often much diversified, and from their incongruous mixture result only harsh discords. On the simplicity and the symetry of musical instruments depend the charms of their mellifluous strains.

Although every kind of oscillation, of which the air is susceptible, must be productive of sound, it does not therefore follow, that such intestine motions are always distinguishable by the human ear. A certain vivacity of impulse, and a certain quickness of succession, seem necessary to affect that organ. It is not only requisite that the tympanum should be struck with some measure of force, but that those strokes should be repeated at short intervals; otherwise their effect will be entirely lost, the first impression being obliterated before the second is made. This property is common to all our sensations, yet appears to belong in a peculiar degree to those of hearing. The capability of the ear to receive the intimations

of

of sound is confined within very moderate limits. It has been estimated, that eight descending octaves form the whole compass of notes which we are able to distinguish. Therefore, reckoning from the top of the scale, after the aërial pulses have become 256 times slower, they are no longer capable of exciting the auditory nerve. How wonderfully superior in susceptibility is the exquisite organ of sight! The eye can discern objects when we may compute their power of illumination to be attenuated many million times. Even vulgar observation discovers the immensity of its range. What a prodigious difference between the fierce rays of a meridian sun and the feeble beams of the silver moon; between the offensive glare of noon-tide day and the faint glimmer of expiring twilight!

If the pulses excited in the atmosphere are at once languid and of slow recurrence, these combined circumstances may be sufficient to extinguish or obliterate their impressions. A series of gentle extended undulations may, therefore, totally escape the cognizance of our sense of hearing. Below a certain pitch of flatness, the human ear

ear seems not adapted to obey the stimulus of sound. Towards the other extremity of the scale, perhaps the utmost elevation of note is only determined by the nature of the elastic medium itself, and the limited celerity or force of percussion. It deserves to be remarked, that deaf persons are affected the most readily with acute sounds, and that, in speaking to them, we generally endeavour to use a sharp, rather than a loud, tone of voice.

I have thus contrasted the different effects produced on the air by a sudden percussion, and by a slow application of force. It is apparent, however, that such distinctions are merely arbitrary, and refer to the ordinary state of our feelings. Nature uniformly proceeds by insensible gradations. In reality, every progressive motion impressed on the subtle fluid which we breathe, is accompanied by a certain vibratory agitation; and conversely, every vibratory agitation implies some degree of progressive motion. From the duration of the primary action results the celerity of progression; from its momentary acceleration is derived the intensity of pulsation. On the former

former depends the breadth, on the latter, the height, of each internal wave. The most steady wind yet betrays its latent undulations, and urges the expanded sails with a reciprocating pressure. When those undulations become excessive, it assumes the destructive character of a gust or squall. If the wind strikes against a sharp edge, it gives birth to a new set of undulations, which, from the suddenness of the shock, are of course narrow and frequent. Hence the mournful whistling of the tempest through the crevices of a door, or among the cordage of a ship.

Several subordinate or interior oscillations in a mass of air, may subsist at the same time. This property belongs, indeed, to every species of matter which is capable of pulsation. When the surface of a pool is agitated, we may perceive the contour of each wave variously marked by vermicular or tremulous motions. A tense chord, if struck, besides the fundamental note, will yield others which this includes. Each single tone of music is in fact composed of at least three natural concords: and from the extension of this principle is derived perhaps the true theory of melody.

After

After the exciting cause has ceased, a sound will for some time prolong its existence. Of the subordinate vibrations, the slower are the first to languish and expire. The flatter tones gradually melt away, while the sharper ones appear in succession to rise with new brilliancy. This curious phænomenon is distinctly observed in a large bell of glass or metal; for it will continue to ring, perhaps, for more than a minute after it has been struck, and the tone which it yields will at intervals mount to a higher and still a higher key, till at last it melts away in a shrill note.

If the air included in a tube, whose farther extremity is closed, be softly struck, the whole column will feel the impression, and will therefore constitute the primary wave. When the tube is open at both ends, the column, being unconfined, will bisect itself, and vibrate on either side from the middle. The notes now produced will consequently be an octave higher than in the former. Both these cases are exemplified in organ-pipes, where the impulse is generated by a slender current thrown against an oblique edge or tongue. If the column of air, however, be struck with unusual

unusual violence, it will not, through its entire length, obey the sudden shock, but will subdivide itself into simple aliquot portions; into halves, thirds, fourths, perhaps sixths or eighths. It is a fact well known to musicians, that, without changing the position of the fingers, one can, merely by blowing strongly into a flute, raise the note at once a fifth, or an octave higher. On this very principle depends the effect of the clarion or French-horn, which brings out the several natural notes, according to the different force with which it is winded. Hence the violent and fatiguing exertion required in blowing it; and hence also the narrow compass and imperfect scale of that martial instrument.

In the case of articulate sounds, the confining of the air does not affect the pitch of voice, but it augments the degree of intonation. The lateral flow being checked, that fugacious medium receives a more condensed and vigorous impulsion. As the breath then escapes more slowly from the mouth, it waits and bears a fuller stroke from the organs of speech. If the lips were much protruded, the human voice would become

become more powerful. A similar effect is produced by a sort of mouth-piece, formed by approaching the palms of the hands; a manœuvre successfully practised by seamen, when their speaking-trumpet is not immediately within reach. But the speaking-trumpet itself is only an extension of the same principle. Its performance does certainly not depend upon any supposed repercussion of sound: repeated echos might divide, but could not augment, the quantity of impulse. In reality, however, neither the shape of the instrument, nor the kind of material of which it is made, seems to be of much consequence. Nor can we admit, that the speaking-trumpet possesses any peculiar power of collecting the sound in one direction; for it is audible distinctly on all sides, and is perhaps not much louder in front comparatively than the simple unassisted voice. The tube, by its length and narrowness, detains the efflux of air, and has the same effect as if it diminished the volubility of that fluid, or increased its density. The organs of articulation strike with concentrated force; and the pulses, so vigorously thus excited, are, from the reflected form of the aperture, finally enabled to escape and spread themselves along the atmosphere. To speak through a trumpet costs a very sensible effort, and soon fatigues and exhausts a person. This observation singularly confirms the justness of the theory which I have now brought forward.

All sounds grow continually feebler as they recede from their source. This diminution of power is explained by supposing, either that the impressions are propagated in radiating lines, or that the intensity of the vibrations relaxes in proportion as these extend and diffuse themselves. The first hypothesis, however, will not bear examination, for it is absolutely incompatible with any regular communication of impulse. Strictly speaking, no range of consecutive particles can possibly compose a right line;* and consequently, if the energy of impression were, at each succeeding step of its transference, to be limited merely to particles that are immediately adjoining, it would soon turn aside and lose itself in a series of intricate mazes. It is only where a cluster, or certain extent, of elementary points mingle their action and balance their mutual irregularities, that motion can be sent

* See Note XXX.

sent in any precise direction. On such combination of individual efforts depends the exactness with which light is reflected from a polished surface.

The other supposition, which refers the transmission of sound to the regular spread of undulations through the atmosphere, must therefore be exclusively adopted. Each wave, or single pulsation, it seems evident, will retain throughout its progress the same absolute momentum or measure of impulse; and consequently, in proportion as it enlarges and expands itself, the corresponding intensity will continually decrease. We may consider the system of aërial undulations as a series of concentric shells, all of equal thickness, and equally charged with motive energies. The degree of those energies, or the scale of alternate contraction and dilatation, must be inversely as the quantity of matter affected in each separate shell, and therefore inversely as the square of the distance from the centre of motion.

That the power of sound decreases continually as it extends through the air, is universally known; but, to ascertain the rate of this diminution by actual observation, seems utterly impossible. The ear, which distinguishes with such nice precision the breadth or duration of sonorous pulses, gives only a vague intimation of their degree of force. We may judge, however, of the law of progression from an analogous fact, which, to facilitate our conceptions, fortunately comes under the fine comprehensive grasp of vision. If I throw a stone into a piece of smooth water, the waves thus formed, at first perhaps of considerable elevation, will gradually subside as they spread around, and will, in approaching the margin, degenerate into an almost imperceptible swell. But these waves must still flatten more slowly than the internal undulations of the atmosphere; because in their propagation, the former only enlarge their circle, while the latter, forming spherical shells, expand themselves in both dimensions.

We are not, however, to conclude, that the pulses excited in an elastic medium invariably spread with equal force in all directions. A notable difference in that respect obtains in the case of sound. If the stroke from which this originates be be not of an expansive nature, but exerted in some particular line, the energy of vibration will be transmitted principally in that direction. The shock of a common explosion, such as that of the blowing up of a powder-mill, is felt all around; while the noise of discharging a cannon is heard farthest in the quarter to which it is pointed. In like manner, the human voice is the most audible immediately in front of the speaker; nor is it merely for the sake of catching the varied expression of features, which add so much to the force of elocution, that we seek to place ourselves opposite to a public orator.

It is the property of all fluids, and the necessary consequence of their internal mobility, that a pressure applied at any spot, is in time uniformly diffused through their whole extent. If air or water, for instance, confined within a vessel, be exposed to the action of a solid piston adapted to an aperture; the portion of fluid immediately contiguous will recede, and admit to a certain degree the entrance of the piston; but the retreat of the whole fluid being prevented by the resistance of the including sides, its component molecules must forcibly approach each other, and, balancing their mutual efforts, suffer a regular condensation, till the repulsion which thence arises forms an equal reaction exerted on every point of the surface of the vessel. This uniform distribution of force takes place, therefore, in the state only of general quiescence, and is the gradual result of certain connected motions or internal derangements. To produce that effect requires some expence of time, which will be the more considerable when the fluid occupies a wide space, or is of a gaseous character, and consequently susceptible of large contractions. If the medium is of an extent unlimited, it is obvious that such equilibrium can never obtain; and, if the exciting impression is likewise slowly made, the adjacent particles, having full time to sympathise with the shock, will form a broad and nearly equable pulse, which will propagate itself regularly on all sides. But, if a sudden blow is given, the effect of the stroke will be confined almost to the particles that are immediately contiguous, and the power or intensity of vibration will be, therefore, principally felt in the direction of the primary impulse.

The

The vigour of those alternate contractions must consequently decrease, as they diverge, on either side, from the axis of motion. We may even venture to assign the ratio of this diminution; for it is a simple proposition in mechanics, derived from the resolution of forces, that the energy of an impression is always proportional to the cosine of the angle of its obliquity. But, from its novelty and importance, the question deserves farther consideration.

Let A, B, and C (fig. 16), be contiguous particles of an elastic fluid in a state of repose or equilibrium: suppose that the particle A receives a sudden impulsion in the direction AB, which carries it forward to *a*; A will then act on B with its acquired repulsion, or with a force proportional to the minute space A*a* of its approach. For the elasticity or mutual repulsion being invariably as the density, this repulsion, in the natural position of the particles A and B, will be denoted by $\frac{1}{AB}$; and by $\frac{1}{aB}$, in their position of derangement. The difference, $\frac{1}{aB} - \frac{1}{AB}$, or $\frac{Aa}{AB \times aB}$, must express the disturbing force, or the impulsive action now

 exerted

exerted on B. But, since *Aa* is extremely small in comparison with AB, we may regard AB $\times$ *a*B as equal to $AB^2$, which is manifestly a constant quantity. Therefore the acquired energy of repulsion between the particles A and B, is proportional simply to *Aa*. But, in advancing to the limit *a*, the particle A also approaches nearer to the particle C: Make C*b* = C*a*, and the very small space of approximation, A*b*, will express the augmented energy which it exerts on C. The elementary triangle A*ba* may be considered as right-angled at *b*; for the angles at the base of the isosceles triangle C*ba* are equal, and the vertical angle AC*a* is evanescent. Consequently *Aa* is to A*b*, that is, the direct action at B is to the oblique action at C, as radius is to the cosine of the inclination CAB.

If, in the position of each of those sentient points, we substitute a cluster of particles, the same consequence, it is obvious, must follow. Rapidity or suddenness of stroke, however, is an essential requisite; otherwise, the more expanded pulsation, mingling its diffusive or lateral energy, would tend in some degree to equalize the diversified

sified powers of radiation. In the production of sound, the primary impulse is besides rarely simple, but compounded of separate efforts, under different divergencies. Hence, though the voice be heard best in front, it is likewise audible in the opposite direction. And, perhaps, one of the advantages of the speaking-trumpet, is to check the natural evagation of the compound stroke caused by the organs of articulation.

Pulsation, vibration, or undulation—these terms we have employed indiscriminately as almost synonimous. Their shades of signification might perhaps express different measures of force. They do not mark any distinct portions of matter, but merely that state or condition of alternate condensation and rarefaction, which the particles, or subordinate systems of particles, throughout a boundless elastic fluid, successively assume. The commotion excited in the general mass seems wholly disproportioned to the feebleness of its primary cause. But this commotion is only an extended series of action and re-action. No permanent force is thereby evolved, and the total momentum, estimated in any direction, continues

continues invariably the same, and equal to the original impulsion. Each particle, in its turn, makes an effort to recede, and again to approach; but these extensive reciprocating movements are likewise accompanied by a regular progressive movement, however small, which, as it spreads from the centre, becomes scarcely discernible. To conceive this more easily, let us imagine a range of particles disposed in a straight line. Suppose the particle A (fig. 17), to receive an impulsion in the direction AE. As it approaches to B, the re-action, or acquired elasticity, of this particle will continually increase, and, at the point *a*, will balance the accelerating force. There, however, A will not stop, but will be carried forward by its momentum to *α*, at an equal distance beyond that limit. It will thenceforth oscillate on either side of *a*, till, gradually wasting its energy, it settles at last in that new position. In its first approach, A will communicate a similar impression to B; the motion of B will next transfer this to C; and thus in regular succession through the whole range of particles. These particles will, therefore, oscillate about the points *a*, *b*, *c*, *d*, *e*, &c. till the imperfect

fect mobility of the fluid extinguishes finally its intestine agitation. Hence, during the time in which the pulsatory influence is conveyed to any given distance, the entire series of included particles will be translated by a progressive movement equal to the minute space, A*a*, of libration.

The same reasoning which applies to single points may be extended to any systems of clustering particles. Though the absolute quantity of progressive motion must remain unaltered, it will yet suffer considerable modification. In travelling from their source the divergent pulsations will continually encounter larger portions of matter, and will, therefore, have their intensity proportionally diminished. Thus, reckoning from the common centre A, the groups of particles at B, C, D, E, &c. will be as 4, 9, 16, 25, &c., and consequently the corresponding displacements B*b*, C*c*, D*d*, E*e*, &c. will be as the fractions $\frac{1}{4}$, $\frac{1}{9}$, $\frac{1}{16}$, $\frac{1}{25}$, &c. Such is the consequence of a single primary impulse. But, if the particle A receives repeated blows either in unison or concord with the vibrations excited, a slow progressive tendency will be communicated to the general mass.

This

This must be the effect of musical instruments on the air.

The undulations of an elastic medium, we have already observed, may be composed of subordinate ones. But undulations of different kinds, and proceeding from different centres, may subsist together. This is plain, from the variety of sounds which can fill the air at the same time. It is likewise confirmed by the phænomena of waves, which cross and traverse each other without the smallest disturbance, on the surface of a sheet of water. I would remark, however, that such irregular pulses must sooner expire than those which are concentric and adapted in unison or concord.

If the mass of recipient fluid be incited by any general motion, this impulsion must evidently have some effect in modifying the celerity or apparent direction of its vibrations. Yet, in most instances, the derangement so produced is scarcely discernible. The atmospheric undulations may serve for an example. It is a pretty strong breeze that travels at the rate of twenty feet in a second, and the rapidity of a hurricane itself has been estimated at eighty feet in a second. Consequently,

quently, the most violent wind could not retard or accelerate the flight of the internal waves, by more than one-fourteenth part. And if it blew right athwart the course of those swift pulsations, the utmost effect of the compound motion would be to cause an apparent deviation of only four degrees. But, in ordinary cases, the variations thus occasioned are much smaller.

Air is the only known body to which perfect elasticity has been ascribed. When it impinges against a firm obstacle, it must consequently resile at an angle equal to that of its stroke; and the same exact equality of incidence and reflection must obviously belong to the aërial pulses which constitute sound. But these propositions, however currently received, ought not to be admitted without examination. To sift the matter fully would require some nice discussion. I shall content myself, therefore, with making a few such remarks as are more directly applicable to the subject.—

It was formerly shown, that every body whatever is within certain limits perfectly elastic; and, that the peculiar effects of collision are determined rather

rather by accidental circumstances, than produced by any intrinsic property. Action and re-action being constantly equal, if the energy developed was confined to the same particles, the celerity of recoil would likewise be equal to that of impact. But, in ordinary cases, during the operation of the repellent force, its influence has time to spread, and of course to attenuate itself. The spot which receives a blow, suffers a proportional compression; and in the act of recovering from which, it involves other adjacent particles, and therefore returns with slower impulsion. Such is more particularly the consequence, when the obstacle struck is of a soft quality. The nature and force of the impinging substance may likewise modify or derange the proper effect. If air, for example, strikes directly against a wall, its particles become approximated, not merely at right angles to the surface, but, by communication, also in some degree laterally: from an effort of distension, therefore, they are sent back in diverging, and not parallel, lines. And this divergency will be the more considerable in proportion to the slowness of impact; because more time will be given for spreading

ing the compression and producing a more regular and equable condensation in the adjacent fluid. If the stroke is oblique the same effect will take place, though in a lower degree. The axis of reflection will make an angle equal to that of incidence; but, there will be, on either side, a profusion of diverging rays. This is experienced when the wind beats against a high wall, for the refluent stream seems to blow in almost every direction. If the motion is rapid however, the reflected wind will be more concentrated. But sound is much swifter than the fleetest wind, and therefore its reflection must be performed with greater accuracy. Still, however, those aërial pulses must acquire, in reflection, a certain measure of divergency or aberration. The parallel rays of approach are darted back in diffusive pencils. A streamlet or pulse of an elastic fluid, is thus reflected from a plane surface, in the same manner apparently, as if this surface had a small degree of convexity. But, in proportion as the velocity is increased, the corresponding curvature regularly diminishes, till at last it becomes evanescent.

evanescent. Hence, on every hypothesis, the rays of light, which shoot through space with a swiftness almost inconceivable, must be reflected with perfect accuracy.

CHAP-

## CHAPTER XIII.

HAVING demonstrated, I trust, satisfactorily that the discharge of heat through the atmosphere is performed by the vehicle of certain aërial pulsations, and having examined at some length, the nature and affections of the undulatory motions which are excited in the body of elastic fluids; it remains for us to apply the principles thus established, to the explication of the various phænomena laid open in the former part of this work. I shall begin with offering only general views, and shall gradually proceed to developе the more abstruse operations.

When heat penetrates, by its own activity, through a solid or inert mass, it successively dilates the several portions of matter which it encounters in its march. In the production of such multiplied displacements, it consumes its expansive energy, and its progress, therefore, is extremely slow. But if those intestine motions are generated by some extraneous cause, the heat, then suffering no impediment to its flight, will passively follow the tide of expansion. And such is the character of atmospheric pulses. The particles of air in immediate proximity to a hot surface, becoming suddenly heated, acquire a corresponding expansion, that propagates itself in an extended chain of undulation; and the minute portion of heat which generated the initial wave, thenceforth accompanies its rapid diffusive sweep. After a momentary pause, a fresh portion of heat is again imparted to the contiguous medium, and the same act is continually repeated at certain regular intervals. The mass of air, without sensibly changing its place, suffers only a slight fluctuation as it successively feels the partial swell; but the heat attached to this state of dilatation is actually transported, and with the swiftness of sound. Nor is the motion of the aërial pulses in any measurable degree retarded by the adhesion of the matter of heat, which is of such extreme tenuity, that, if not detained and cramped by the *inertia* of other bodies, the smallest possible force is sufficient to impel it with a celerity yet much inferior to that of light.

The same principle will likewise explain the dispersion of cold. For the atmospheric particles that come in contact with a cold surface, must suffer a sudden contraction, which will shoot its vibratory influence through the general mass: and the cold wave thus excited will, in its spreading tremulous flight, still retain the same distinctive character. Each of the minute parcels of air, as they successively feel a contractile disposition, will suffer a corresponding depression of temperature, or will permit a certain part of their heat to escape. The heat so liberated, is again instantly absorbed by the portion of air next behind, which, having contracted, is now recovering its tone. Though the motion of the aërial pulses, therefore, is the same as in the former case, yet the direction of the subtle element of heat is exactly reversed. Heat is, with the rapidity of sound, conveyed from all quarters to the cold surface, as to a common centre.

These internal waves, whether of the quality of hot or cold, must evidently have all the properties which belong to elastic pulsations. Their motion is not apparently deranged by any mechanical agitation of the atmosphere: and it was found, that the blowing strongly with a pair of bellows across the direction of the undulatory current, between the canister and the reflector, did in no perceptible degree affect the action or the focal ball. Each wave, or hemispherical shell, through the whole of its expansive sweep, retains the same absolute excess or defect of heat. But the intensity of this difference, or the partial elevation or depression of temperature, diminishing, therefore, in proportion as they spread, must, as in the case of radiations, be inversely as the square of the distance from its source. It is not equal, however, in all directions; at right angles to the exciting surface, the power is greatest, and regularly declines on either side as the cosine of obliquity. The shell of aërial pulsation, it was shown, is not uniformly condensed or dilated, but after the law now stated: and these theoretical conclusions were abundantly confirmed by experiment. Nor will the force or character of the undulations be altered in any respect, by traversing air of a very different or irregular temperature. Each distinct portion of that medium, being

being successively affected with a disposition to expand or contract, will likewise, at the same moment, assume the appropriate excess or defect of heat. A wave, for instance, that is originally hot, will always be hotter than the mass of fluid through which it travels: in fact, it will only superadd, in its passage, a certain measure of dilatation or of heat; and whether it encounters hot or cold streams, it will preserve the same relative excess of temperature. This deduction was entirely consonant to observation: for, having placed the canister and the reflector upon two tables a little separate, and holding a red-hot poker stretched across, and somewhat below the vacant space, the effect on the focal ball was not thereby at all changed; nor could the smallest alteration be perceived, when a block of ice was suspended above the course of the aërial waves to the reflector.

Those waves, therefore, spread without interruption or modification of any kind from the state of the intervening fluid. But when, in their progress, they strike against a firm obstacle, they undergo a very material change. This obstacle produces an effect contrary, yet analogous, to that of the exciting surface; for, absorbing more or less the heat of the impinging wave, it diminishes proportionally the measure of intensity or rarefaction; and the wave, so enfeebled, next suffers reflection. If the reflecting surface is an exact plane, the hot pulses will preserve the same mutual divergency; but if it has a suitable concavity, they will tend to some focus, and consequently will again converge and unite their accumulated power. In thus concentrating themselves, their heat or dilatation, collected into a narrow space, must have its intensity, or its temperature, in a corresponding degree augmented. But the reflection of those pulses is not performed with geometrical accuracy; it is affected by a certain small aberration, arising, as was shown, from the limited velocity of sound. And such result accords perfectly with observation. I need scarcely remark, that the same mode of argument will conversely apply to the partial absorption, and the subsequent reflection, of cold pulses.

The particles of air contiguous to a hot surface must evidently receive the same charge of heat, and

and this acquired temperature will determine the power of their remote pulses. Accordingly, it was found, that the impression made upon the focal ball is always proportional to the difference between the temperature of the canister and that of the surrounding atmosphere. When a pulse is once excited; the agitation will continue for some time after its cause has ceased to operate. Hence we are not warranted to conclude, that all the consecutive pulses are efficacious. The hot waves may succeed each other, at greater or shorter intervals, according to circumstances. Their formation depends on the sympathetic energy of the primary conterminous surface; for all communication of heat is necessarily preceded by some vibratory movement. If the air is divided from the hot mass by a wide limit, the prelusive pulse will be comparatively slower. The disposition of the exciting surface to impart its heat, must evidently be more languid in proportion to the distance, however minute, of that receptive fluid: and, in every case where the same force is exerted, the time of a pulsation is directly as the space lying between the extreme affected

 points.

points. Thus, the surface of the canister may deposit its heat as fast almost as the pulses would succeed each other in the surrounding air ; or it may not be in a condition to make those deposits but at considerable intervals. Hence the tremulous rays excited in the elastic medium, will either have each consecutive pulse charged with heat, or only certain favoured pulses which follow at some regular distances.

Such appears to be the mode of operation by which a vitreous and a metallic surface produce their very different effects in discharging heat. This curious fact was brought out at an early stage of our inquiry, and we have since had frequent occasion to refer to it as involving an important principle. It was observed, in general, that those surfaces owe their distinct qualities, with respect to heat, to the different measures of their approach to the contiguous atmosphere. Physical contact implies a finite interval of division, and is therefore susceptible of various degrees of approximation. When no extraneous power is applied, the proximity of the bounding surfaces will be determined merely by their affinity or mutual

mutual attraction. But the strong affinity of air to glass is shown by a variety of facts. A glass vessel, after having been cracked, is yet often perfectly air-tight, a property not observed under similar circumstances in one of metal. A thin film of air insinuates itself into the crevice, from which it cannot be dislodged without introducing some liquid which has a still superior attraction for glass. In like manner does air seem to adhere obstinately to the inside of a barometric tube; and hence the necessity of boiling the included mercury, in order to expel the latent fluid. Glass exhibits a remarkable power of abstracting moisture from the air, before it is completely humid, or has attained the point of saturation : and that property argues the very close proximity of air to the surface of the confining glass, else this could not exert such a prepollent force of attraction. But the inference which we draw is capable of a more direct and conclusive proof. It was found, that a metallic surface, when striated or covered with fine parallel furrows, has its power of discharging heat more than doubled. An obvious consequence, however, of this change, is the

the production of a closer, though a partial contact with the atmosphere. The multiplied slender ridges, protruding themselves beyond the ordinary limit, must obtain an approximation to the encircling medium, analogous to that of glass. Nor will the effect of this contracted proximity be counterbalanced by the increased remoteness of the corresponding furrows. The contact of air with glass, and that with metal, seem to occupy the two extremes: in the former, the action is scarcely augmented by a nearer approach; and in the latter, it is not sensibly diminished by enlarging the distance. The power of glass to emit heat, appears nowise altered by having its surface filled with regular scratches. To conceive the effect of that process on a plate of metal, let us suppose those raised flutings to acquire only half the energy of glass, or four times that of a smooth metallic surface. The one-half of the striated surface, being composed of prominences, will, therefore, have an action equal to $4 \times \frac{1}{2}$, or 2, and the other half, consisting of similar cavities, will have an action less than one-half, and which may be reckoned at one-third, or one-fourth. Thus the compound

compound result is $2\frac{1}{3}$ or $2\frac{1}{4}$, being rather more than double the simple effect of a smooth metallic surface. But if the quantity of protuberant matter be diminished, its peculiar energy will become again enfeebled or destroyed. Thus, when a striated metallic surface is likewise furrowed across, its power to discharge heat is almost the same as if it were wholly smooth.

This reasoning appears abundantly satisfactory; but it receives ample confirmation from the experiments made on thin plates, which give a singular precision to our ideas on the subject. Tin-foil, whose thickness exceeds not the 600th part of an inch, attached to the glass side of the canister, was found to discharge heat with the same power as the mere tin surface. When a coat of silver leaf, only the 150,000th part of an inch in thickness, was applied however to the glass, that power seemed to be, though in a very small degree, augmented. The action of the subjacent vitreous surface was therefore scarcely felt at the distance of the 150,000th, but did not at all penetrate so far as the 600th part of an inch. Yet with successive coatings of isinglass, from the millionth to the

the 200th part of an inch in thickness, applied to the tin side of the canister, a series of regular increasing effects was traced. The energy of a metallic surface has consequently such enlarged limits, and extends at least to the 200th part of an inch. By its repulsive force, the atmospheric boundary is made to retire before the coating. The distance of that boundary from the external surface must be, therefore, determined by the excess of the metallic limit above the thickness of the isinglass. As this coating increases in thickness, it continually approaches nearer to the conterminous air, and, with relation to heat, it hence partakes more and more of the quality of glass or pigment. In every case, the power of a surface either to discharge or absorb heat, seems to be inversely as its width of separation from the contiguous medium.

But the views now stated will be rendered more intelligible by the help of diagrams. The comparative approximations of metal and glass to the air that bounds them, are represented in figures 18 and 19; in which AB denotes the hot or cold surface, and C D its atmospheric limit.

The

The vacant interval is, for the sake of distinctness, magnified 100 times, and the other figures are delineated after the same enlarged scale. Figures 20 and 21 exhibit coats of isinglass applied to metallic surfaces, where C D, as before, represents the confines of the air, and E F the exterior surface of the isinglass, which, from its greater thickness, makes a nearer approach to that boundary in fig. 21 than in fig. 20. Figure 22 expresses a striated or fluted surface of metal. If a plane be supposed to bisect those prominences, or rather to pass through their centre of gravity, the atmospheric limit C D will recede from this to its proper distance. The interval that divides C D from the base A B, will therefore be somewhat increased; but the protuberant parts of E F will obtain a close and artificial approximation to the contiguous air. If these asperities are unusually sharp, they may even penetrate so far as to disturb the atmospheric boundary, and give it, in some measure, an undulous contour. And such must always be the effect of a vitreous surface when striated; for, in this case, the action being confined within a very narrow space, the more distant

distant parts of the glass have no influence to balance the inequalities. Figure 23 denotes a surface of that kind, and the waved line C D expresses the contiguous atmospheric limit. Thus no actual approach is now made to the surrounding medium, and consequently no alteration of power is produced.

The various disposition of bodies to emit heat is, therefore, derived from the diversified quality of their contact with the surrounding air. When there is an intimate approach of the adjacent boundaries, the fits of emission succeed each other with rapidity; but if the space of separation is considerable, they will follow slowly, and at great intervals. A hot surface of glass makes its deposits on the contiguous layer of atmosphere about eight times faster than one of metal: and the sums of these successive decrements, or the total quantities of heat discharged by pulsation, must consequently have that relative proportion. But the same principle explains likewise the various aptitude of different substances to receive or absorb heat from the impinging aerial waves. A surface of metal is in a condition to sympathise with

with the successive appulses eight times seldomer than one of glass. Its receptive power, therefore, acts only during one eighth part of the time in which that of the latter is exerted. Hence, while a surface of glass or paper absorbs almost the whole of the incident heat, a metallic surface does not scarcely detain the eighth part of it, but suffers the remainder to be reflected. Hence, too, the influence of a coat of isinglass applied to a speculum, in diminishing the quantity of reflection; for it occasions an approximation of the atmospheric boundary, and consequently augments the previous absorption of heat. These films were found, according to their several degrees of thickness, to produce an extended series of effects. If a speculum has its surface filled with parallel scratches, they not only disturb somewhat the regularity of reflection, but diminish its power by their closer proximity to the contiguous air.

Yet we are not to conclude that any surface, whether it receives or discharges heat, acts by a simultaneous impression over the whole extent of its atmospheric boundary. If the particles of this contiguous

contiguous stratum were all equally and at once affected, they would, by their joint influence, produce only a general pulsation directly forwards, without any degree of divergency or deviation whatever, except near the edges. In the next range of air, each particle, being urged by a combination of oblique forces, proceeding from numerous collateral points, would acquire a regular impulsion in the line perpendicular to their plane. The same effect would be repeated in every successive layer; and consequently this supposition is quite incompatible with the diffusive radiation which experiment shows actually to obtain. The phænomena require us to admit a multiplicity of distinct centres of undulation. These must be disparted at certain comparatively wide intervals over the atmospheric boundary. Each initial particle then will act on the cluster of points immediately before it, and excite those various and dispersive pulses which are found to take place. Nor will the impressions, springing from such divided sources, at all derange or modify each other's effects. It is not difficult to comprehend this statement, but fig. 24. exhibits it it to the senses. The circular space represents the bounding stratum of air, and the numerous dotes arranged at equal distances, denote the salient molecules, or the centres of vibratory agitation.

There are even data sufficient for assigning the proportion of those energetic molecules to the expanse of atmospheric surface in which they are disposed. It was already noticed, and will be fully proved in the sequel, that the waste of heat by pulsation from a painted canister is the same as if it were transported by the successive encircling shells of air, raised constantly to that temperature, and receding with a velocity of one foot in about six minutes. But this heat is actually conveyed away with the swiftness of sound, or at the rate of 1142 feet in a second. Each vibrating ray, however, is only charged through half its length; for, in every species of waves, the depressions are equal to the corresponding elevations. With the half of that velocity, or 571 feet in a second, it would therefore have produced the same effect, if it were uniformly heated. Hence the dispersed physical points, by which only the heat is communicated, or from which it is received,—*rari nantes in gurgite vasto!*—must be to the whole stratum of air in which they float, in the ratio of 1 to 360 × 571, or must constitute only the 205,560th part of that extended film.

If we regard those centres of pulsation, or energetic molecules, as only minute circles, it will be easy to determine the proportion of their diameters to their mutual distances. Being arranged with perfect symmetry, their lines of junction must evidently form a system of equilateral triangles. Each angle will be occupied by a sextant; and since three sextants compose a semicircle, the elementary semicircle D E F (see fig. 25.) must amount to the 205,560th part of the correlative equilateral triangle A C B. But the surface of this triangle is equal to $\frac{1}{2} \surd \frac{3}{4} \times AB^2$; and the area of the semicircle D E F is equal to $\frac{1}{2} \pi \times AD^2$, $\pi$ denoting the ratio of the diameter to its circumference. Consequently, the semicircle is to the equilateral triangle as $\frac{1}{2} \pi \times AD^2$ is to $\frac{1}{2} \surd \frac{3}{4} \times AB^2$, or as $AD^2$ is to $\frac{1}{\pi} \surd \frac{3}{4} \times AB^2$. Whence $\frac{1}{\pi} \surd \frac{3}{4} \times \frac{AB^3}{AD^2} = 205{,}560$, and

AB²

$\frac{AB^2}{AD^2} = \pi \surd \frac{4}{3} \times 205{,}560 = 745{,}689$, and therefore $\frac{AB'}{AD} = \sqrt{745{,}689} = 863\frac{1}{2}$. Thus, the radius of each pulsatory circle amounts not to the 800th part of their mutual distance.

Such, then, is the extreme limit to which the hot or cold vibrating molecules of our atmosphere seem capable of being condensed. The degree of their condensation has no connexion whatever with the proximity of the exciting or absorbing surface: it proceeds undoubtedly from the nature of the fluid medium, and may differ materially in each distinct species of gas. To trace its intimate cause, indeed, is perhaps no more possible than to discover the sources of some of the most ordinary properties of bodies. But although those efficient particles are capable only of a certain degree of mutual approximation, they will admit of an unlimited distension. In propagating their energy, the rays of vibration continually spread; nor is their divergency ever compensated by any subsequent concentration. The density of the hot particles collected even in the focus of the speculum, must be always inferior to that which

 was

was originally impressed at the surface of the fronting canister. Hence, in the appulses of heat, the affected molecules will be sufficiently disparted, and the contiguous absorbing surface will of course exert its full influence.

CHAP.

## CHAPTER XIV.

I HAVE now discussed at considerable length the mode of operation by which heat is discharged or communicated through the agency of aërial pulsation. I have endeavoured to estimate the different degrees of contact or mutual proximity which obtain in nature, and to ascertain their relative influence in moulding that abstruse and singular process. One leading principle is made to connect the various facts; and the perfect agreement of theory with observation seems to confirm its justness in the completest manner. Yet I will not dissemble that the explication advanced requires some stretch of thought to comprehend it rightly, and, although consistent in all its parts, it rather appears to savour of excessive subtlety and refinement. Such, however, is necessarily the character of every strict inquiry into the properties of the internal motions of fluids. The imagination, unassisted by the direct appeal of sight or touch, is apt to be fatigued and bewildered by their multiplicity and unavoidable complication.

But the subject of the propagation of heat through the atmosphere is not yet exhausted. A variety of ulterior questions still present themselves to our research. Is aërial pulsation the only mean by which a substance disperses its heat? If some other mode besides exists, is air in this case likewise the active instrument? Is that dissipation at all accelerated by the mechanical agitation of the surrounding fluid? And does its efficacy in any degree depend on the nature or qualities of the hot surface? If a body lost its heat by the single agency of atmospheric pulses, no alteration of effect could result from the play, however rapid, of the circumfluent medium. Yet there is scarcely an observation more familiar than the remarkable influence of currents of air in cooling a surface over which they pass. On a larger scale, the action, in that respect, of furious sweeping winds is prodigious. We may, therefore, safely infer, in addition to the species of dispersion already investigated, the existence of of another mode of dissipation, produced by the frequent renewal of the contiguous air, of which the several portions, as they successively apply themselves to the surface, imbibe, and then carry off its excess of heat. But to develope fully the nature and conditions of this secondary process, demands some closer examination.

Though the power of abstracting heat from a body may continue the same, its effect will regularly diminish. The portion of heat which is at each succeeding instant transferred, must evidently bear a relation to the whole quantity, or to the excess of temperature above that of the surrounding fluid. But this variable waste or abstraction can only be deduced from the actual measures of heat, or the corresponding temperatures observed at certain limited periods. The problem is of the same kind precisely as one that occurs in dynamics, and which, in most cases, admits of an easy solution; namely, to determine, at any interval of time, the velocity of a point whose motion is, after some given law, continually retarded. It is assumed as a general principle, that the decrements of heat are proportional to the difference of temperature of the conterminous surfaces. On this supposition, the successive temperatures of a substance exposed to cool, would, at equal periods, form a descending geometrical progression; for it is the peculiar property of such a series, that the difference between any term and the one next to it, is always proportional to the term itself. Thus, if a body lost every minute the hundredth part of its heat, or rather its excess of temperature above that of the surrounding atmosphere; at the end of the first minute, it would still contain the $\frac{99}{100}$th part of the whole; at the end of the second minute, the $\frac{99}{100}$th part of this, or $(\frac{99}{100})^2$; at the end of the third minute, only $(\frac{99}{100})^3$; and so extended with the several powers of the radical fraction. Consequently, the time elapsed, or the interval between any two terms, is always proportional to the difference of their indexes, or that of their logarithms. If, in this instance, the body had continued to discharge its heat with the same profusion as at first, the whole would have been spent in the space of 100 minutes. This number is the reciprocal of $\frac{1}{100}$, which expresses the *rate*

of

of cooling, or the frigorific energy: I shall, for want of a better term, therefore, denominate it the *range* of cooling. If the body had cooled twice as fast as we have now supposed, its temperature, at the lapse of each minute, would have been represented by the series $\frac{49}{50}$, $(\frac{49}{50})^2$, $(\frac{49}{50})^3$, &c. But if it had cooled uniformly after that rate, the temperature sinking successively to $\frac{48}{50}$, $\frac{47}{50}$, $\frac{46}{50}$, &c., it must have reached the limit of equilibrium in 50 minutes, which consequently denotes the range. This range of cooling has therefore the same ratio to the interval of time between any two temperatures, that unit has to the difference of their corresponding logarithms on the Neperian scale, or that .4342945, the *modulus* of the system, has to the difference of their common logarithms. The anti-logarithm, or the number corresponding to the logarithm, of the *modulus*, or the *modular ratio*, is 2.7182818, whose reciprocal, or .3678794, must express the temperature at which the body will arrive during the range of cooling, or in the time that would have been required for the dissipation of its whole heat, if it had descended by a simple arithmetical progression.

progression. It hence appears, that more than one-third of the heat still remains after the time due to an uniform series is elapsed; or if, by applying the doctrine of *continued fractions*, we convert the *modular ratio* into its approximate values, we shall find that $\frac{71}{193}$, or still more accurately, $\frac{2073}{5635}$ will denote the remaining portion of heat after the expiration of that full period. A ready method, therefore, offers itself for discovering the rate of cooling, and without any computation: it is, to watch the time when the temperature is reduced to the $\frac{71}{193}$th part of the whole. But the comparative degree of frigorific energy is conveniently inferred, from the time wherein one-half of the heat is lost, and which may be found either by observation, or by an easy logarithmic analogy. This latter mode I very frequently used.

A geometrical progression, though perpetually diminishing, is never extinguished; and therefore it is impossible for a body once heated, after any lapse of time, absolutely to regain its equilibrium. Thus, though the range of cooling has been quintupled, the $\frac{1}{148}$th part of the heat yet remains;

remains; and even after it is repeated ten times, there still subsists a residue equal to the $\frac{1}{22025}$th part of the whole. Experience confirms in general the correctness of this conclusion. But it was remarked formerly, that infinite divisibility belongs only to mathematical conceptions, and is utterly excluded from the physical world. All the changes in nature are accomplished by certain successive steps.

The preceding conclusions evidently rest on the accuracy of the fundamental principle, that the decrements of heat are uniformly proportional to the corresponding temperatures. This is certainly true in the communication of heat among solid bodies, and in its dispersion by the vehicle of aërial pulses. Whether it holds equally in all cases, must be the subject of patient inquiry. If the frigorific energy proves to be variable, it will be requisite to observe the variations of temperature at short intervals. According to circumstances, I noted the quantities with different frequency: at every two minutes, perhaps, or only after the space of an hour.—These remarks being premised, I now resume the experimental investigation.

## EXPERIMENT XLIII.

In a close room without a fire, having placed a thin hollow globe of planished tin, four inches in diameter, and with a narrow neck, on a slender metal frame or stool, and resting against the sharp edge, I filled it with warm water, and inserted a thermometer. The air of the room was perfectly steady, and at the temperature of 15 degrees centigrade. I noticed carefully the progress of the ball in cooling: from the station of 35°, till the internal thermometer sunk to the middle point, or 25°, the time elapsed was 156 minutes. I next painted the surface of the ball with a coat of lamp-black, and again filling it with warm water, scrupulously repeated the experiment. The same effect was now produced, or one-half of the heat expended, in the space of only 81 minutes.

Augmenting those results in the proportion of the logarithm of 2 to the *modulus* of the system, or of 70 to 101, we deduce the range of cooling, or the time in which the globular mass would have

have spent its whole charge of heat, if it had continued to suffer the same uniform diminution as at first. With the metallic surface that limit was therefore 225 minutes, and with the painted surface 117 minutes; in other words, the former must have lost every minute the 225th part of its heat, and the latter the 117th part only. But if the heat were abstracted by the single action of aërial pulses, the metallic surface ought of course to cool eight times slower than the painted one, or expend only the 936th part of its charge each minute. Its actual rate of cooling, however, is more than four times greater, and therefore some other cause must necessarily have joined its influence to accelerate the dissipation. Whether this auxiliary power is exerted equally on either kind of surface, or in what proportions it distributes itself, must be discovered by other means.

But the experiment, considered in itself, claims our attention not less than by the consequences which it involves. The application of a coat of pigment to a metallic surface, instead of retarding the effect, almost doubles its discharge of heat. This fact, equally curious and important, is most

contrary

contrary to the prevalent notions, and seems not to have been hitherto observed. Had the reverse taken place, we should have readily satisfied ourselves with attributing it to the slow conducting quality of the superficial crust, which might obstruct the passage of heat to the encircling atmosphere. In reality, the coat of lamp-black has some influence to impede the process of cooling, yet comparatively in a very small degree. Nay, a tin canister, filled with hot water, will cool considerably faster after it is covered with flannel, and would require the farther addition of one or two folds to make it cool at the same rate as before; the profuse energy of its unmetallic surface being then compensated by the retardation arising from the thickness of the spongy mass. These remarkable facts bid defiance to the sort of loose philosophy, which, without requiring any exertion of mind, pretends to explain every thing. They teach us the necessity of a sober scepticism, and demonstrate the great utility of extending a critical inquiry into the various popular branches of physics.

## EXPERIMENT XLIV.

Another similar ball of tin being provided, exactly of the same size as the former, I filled each with warm water, inserted its thermometer, and exposed them both together on their slender stools out of doors to the action of the wind. As I was anxious to preclude every circumstance which might disturb the accuracy of the results, I always chose, for the time of making the observation, the approach of evening, when the light reflected from the sky had become so much enfeebled as no longer to produce any sensible calorific effect. From the moment when the thermometer in either ball stood at 20° above the temperature of the atmosphere, until it bisected that interval, or had sunk 10 degrees, the time was carefully measured. In a gentle gale, the clear ball was found to lose half its heat in 44′, and the painted one in the space of 35′. In a pretty strong breeze, the times elapsed were respectively 23′, and 20¼′. But exposed to a vehement wind, the times required were only 9½′ and 9′.

It

It is indubitable, therefore, that the quick renewal of the contact of air has a most decided power to accelerate the cooling of a body. The effect is likewise apparently the same upon any kind of surface; for, in proportion as the change of atmosphere is more rapid, and its influence of course predominates, the rate of cooling approaches to equality. Thus, under the action of a very high wind, the difference in that respect between the clear and the painted ball did not exceed the twentieth part of the whole.

But those several results deserve a nearer inspection. In the first observation, the ranges of cooling that belong to the two balls are easily computed to be $63\frac{1}{2}'$ and $50\frac{1}{2}'$. But in a close room, the ranges were respectively 225 and 117. Hence, on the clear ball, the gentle current of air must have exerted an influence expressed by the difference of the two fractions $\frac{1}{63\frac{1}{2}}$ and $\frac{1}{225}$, or it must every minute have occasioned an additional loss of the $88\frac{1}{2}$th part of its whole heat. On the painted ball the auxiliary effect of that current, or $\frac{1}{50\frac{1}{2}} - \frac{1}{117}$, amounts to the 88.8th part,

part, being very nearly the same as the last. And if we pursue the comparison through the other examples, we shall find the frigorific influence of the wind surprizingly equal on both balls. Thus, in the second observation, the ranges of cooling are 33′.2 and 29′.2; but $\frac{1}{33.2} - \frac{1}{225} = \frac{1}{38.9}$, and $\frac{1}{29.2} - \frac{1}{117} = \frac{1}{38.9}$. In the extreme case, the ranges were only 13′.7 and 13′; and consequently the action of the high wind on the clear ball was $= \frac{1}{13.7} - \frac{1}{225} = \frac{1}{14.6}$, and that on the painted ball was $= \frac{1}{13} - \frac{1}{117} = \frac{1}{14.6}$.

The contact of fresh portions of air renewed in quick succession, not only, therefore, accelerates remarkably the cooling of bodies, but has, in that process, exactly the same measure of effect on a surface of metal, as on one of pigment. These two kinds of surfaces are most opposite in their relation to atmospheric pulses, and it is hence most reasonable to presume, that the same equality of influence would be exerted on every surface of an intermediate species. This inference will be fully confirmed in the sequel. But the agitation of the atmosphere, on which alone de-

 pends

pends its accelerating effect, might also be produced by other causes besides mechanical impulsion. It seems natural to suppose, that a hot body itself will, in the mere act of cooling, occasion an internal motion, however partial or confined, among the particles of air situate within its immediate vicinity. The action derived from that source, and which is necessarily the same whether the surface consists of metal or pigment, being added to the unequal impressions of the atmospheric pulses, their combined energies must obviously approach nearer to a ratio of equality. And, should the auxiliary effects increase in the higher temperatures, when the heat must have more efficacy to excite the surrounding air, this conjecture will be converted into certainty.

## EXPERIMENT XLV.

I filled the same two balls with boiling water, and set them to cool as before, in a close room. After the thermometer came to stand in each at 92°, or 80 degrees above the temperature of the surrounding air, I observed the time which it required to sink 10 degrees. With the clear ball, that

that interval was 15½ minutes; but, with the painted ball, it was only 10½ minutes.

These results being augmented in the proportion of the difference of the logarithms of 80 and 70, the compared temperatures, to the *modulus* of the system, will give the ranges of cooling; which are, therefore, respectively 116′ and 79. Hence the process of refrigeration has been much accelerated in both balls: in the clear ball the rate is almost doubled; and in the painted one, it is augmented only by one-half. The heat has, therefore, an effect similar to that of a gentle current of air, in giving to different surfaces a tendency to equalize their progress in cooling. Thus, at a low temperature, the respective rates of the clear and the painted ball were as 13 to 25; but, at a high temperature, they were as 13 to 19.

But we may now distinguish the elements of the compounded energy, or develope the separate influence which the atmosphere exerts, by touch, and by pulsation. The fractions $\frac{1}{225}$ and $\frac{1}{117}$ denote the respective rates of cooling between the temperatures of 20° and 10°: their difference, or $\frac{1}{244}$, will consequently express the difference between the pulsatory powers that belong to the metallic and the painted surfaces. The seventh part of that quantity, or $\frac{1}{1708}$, must therefore denote the influence of a surface of metal, and eight times this again, or $\frac{1}{213\frac{1}{2}}$, must denote the influence of a surface of pigment. Hence we derive the effect due to the repeated contact of air; for it is obviously equal to $\frac{1}{225} - \frac{1}{1708} = \frac{1}{259}$; and the same result will be obtained from $\frac{1}{117} - \frac{1}{213\frac{1}{2}}$. Between the temperatures of 80° and 70°, we have $\frac{1}{116} - \frac{1}{1708} = \frac{1}{124\frac{1}{2}}$, for the effect of atmospheric touch; or, $\frac{1}{79} - \frac{1}{213\frac{1}{2}} = \frac{1}{125}$, for the same, being a coincidence sufficiently near. Thus, it appears, that at low temperatures the portion of heat dissipated from a painted surface by the repetition of aerial contact is somewhat less, and in high temperatures considerably greater, than what is spent by pulsation.

These deductions are farther elucidated and confirmed by what takes place on the immersion of of hot bodies in liquids. It was proved, that the discharge of heat by pulsation is exclusively the property of gaseous fluids: accordingly the peculiar nature of the surface which is plunged into water, shows no effect whatever to accelerate the rapidity of its cooling. On immersing the clear and the painted ball successively, I found that they lost their heat exactly with the same degree of facility. In like manner, a large glass ball cooled with the same progress, if plunged, either naked or encased with tin-foil. Nor does the rate of cooling continue uniform, but diminishes very perceptibly as the temperature is depressed. This fact had occurred to me before the experiments now related were made. With a view to discover the power with which water conducts heat, I fixed a differential thermometer, having its sentient ball of black glass, in a vertical position within a large transparent vessel, which I filled with cold water to the height of two or three inches above the instrument, and exposed the whole apparatus to the action of a bright sun. The coloured liquor soon mounted several degrees; but, on decanting off the water,

 and

and refilling the vessel with other water as hot as the differential thermometer could bear, there was still a rise, though much smaller than at first. It appeared evident, therefore, that the heat incessantly communicated by the solar rays to the black ball is faster absorbed or consumed by hot, than by cold, water. But the same conclusion was derived in another way.

## EXPERIMENT XLVI.

Employing a stop-watch with seconds to measure the time, I plunged the clear four inch ball as quickly as possible into the centre of a large mass of water, reduced to the point of congelation. In the space of 24 seconds, the inserted thermometer sank from 90° to 70°; but 138 seconds elapsed while it sank from 9° to 7°.

These limits of temperature are proportional, and consequently, had the same frigorific energy obtained in both, the corresponding intervals of time must have been likewise equal. But the power of water to abstract heat appears to become much feebler as its temperature is lowered,

and

and even to suffer a greater diminution in that respect than air itself. The acceleration of effect in high temperatures is undoubtedly produced by the increased internal motion occasioned by the greater degree of expansion which is then communicated to the adjacent portions of liquid. A stream directed against the hot ball has a similar action. Thus, it will cool faster if I drag it through a piece of water; and more so, when I pull it along with force.

But in the case of a hot body exposed to the influence of a current of air, it is of importance to discover the precise relation that connects the power of cooling with the celerity of impact. For this purpose, it would be requisite to hurl the mass, which is subjected to trial, with an equable motion through the atmosphere, and with different degrees of velocity. Not having an opportunity, however, of appropriate machinery, I was obliged to content myself with a readier, though more imperfect, method. I fastened a long cord to the neck of the tin globe, and having filled it with boiling water, I whirled it for several minutes round my head, counting the number of revolutions, and marking, by means of a stop-watch, the time elapsed. It was thence easy to compute the velocity; and I could increase this at pleasure, by lengthening out the cord, and applying more force. This exercise proved surprizingly fatiguing, and when my utmost strength was employed, I could not maintain the exertion steadily for more than two minutes: but during that short space, the ball had travelled near a mile and a half. Comparing the logarithm of the ratio of the temperatures immediately before and after this flight with the time which intervened, I calculated, in each instance, the corresponding range of cooling. The results thus obtained I would not esteem rigorously accurate; they may, however, be considered as near approximations to the truth, and as sufficient for the establishing of any general conclusions. I shall select a few distant terms for a specimen, being the mean, in round numbers, of various repeated trials.

EXPERI-

## EXPERIMENT XLVII.

In a calm evening, the clear four inch ball, suspended out of doors, cooled with the range of 120. Holding it finely by the neck with my hand stretched out, I walked in a wide circuit, at a smart equal pace of 400 feet in the minute; the range was now 60. Then, attaching the cord, and whirling it about my head with the velocity of 20 feet *per* second, I found the range diminished to 30. Extending the cord, and making a vigorous exertion, I gave the ball a rapidity of 60 feet *per* second; and, with this extreme celerity, the range of cooling was at last reduced to 12 only.

It is plain, from inspection, that the ball had its velocity, successively tripled; and comparing the respective rate of cooling, it was doubled at the second trial, and at the third increased two times and a half; thus evidently marking a tendency to follow the proportion of the velocity itself. But this correspondence will clearly appear, if we deduct the ordinary influence of cooling. The separate action of the atmospheric current is

$\frac{1}{60}$

$\frac{1}{60} - \frac{1}{120}$ or $\frac{1}{120}$. When the velocity was 400 feet each minute, or $6\frac{2}{3}$ feet *per* second, it is $\frac{1}{60} - \frac{1}{120}$, or $\frac{1}{120}$. With the velocity of 20 feet *per* second, that action was $\frac{1}{30} - \frac{1}{120} = \frac{1}{40}$. And when the velocity amounted to 60 feet *per* second, the distinct effect of the artificial wind was $\frac{1}{12} - \frac{1}{120}$, or $\frac{1}{13\frac{1}{3}}$. But these fractions $\frac{1}{120}$, $\frac{1}{40}$, and $\frac{1}{13\frac{1}{3}}$, obviously rise by successive tripling.

The refrigerant power of a stream of air is, therefore, exactly proportional to its velocity. Hence we may determine the rate of cooling that corresponds to any given velocity of the ball. Let $v$ denote that velocity in feet *per* second, then the fraction $\frac{6\frac{2}{3} + v}{800}$ will express the rate of cooling; or its reciprocal, or $\frac{2400}{20+3v}$ will express the range. For example, if the velocity be $26\frac{2}{3}$ feet *per* second, then $\frac{2400}{20 + 80} = 24$, and consequently the ball, carried through the air at such a rate, would lose the 24th part of its heat every minute. But the *formula* will apply to any other body, the ordinary

ordinary range of cooling being ascertained by observation, and multiplied by 20 to give the numerator. Suppose this ordinary range $= T$, and the range corresponding to any velocity $= t$; then in general $20 + 3v : 20 :: T : t$, and $3v : 20 :: T—t : t$, and therefore $v = \frac{20}{3}\left(\frac{T—t}{t}\right)$, in feet *per* second; or $v = \frac{T—t}{t} \times 4\frac{1}{2}$, in miles *per* hour. Instead of the ranges of cooling, denoted by T and $t$, the times in which a body loses an aliquot part of its heat, as the half or the third, may be substituted, since the latter are always proportional to the former. The *formula* may likewise be reverted: put $V =$ the velocity of the wind in miles *per* hour, and we have $\frac{T}{t} = 1 + \frac{V}{4\frac{1}{2}}$ to express the relative degree of cooling.

These *formulæ* may be conveniently exhibited by geometrical figures. Draw the perpendicular BC = AB (fig. 26), join AC, and produce it indefinitely. If BE express the velocity of the wind, AB denoting that of $4\frac{1}{2}$ miles *per* hour, the perpendicular DE will express the corresponding increased rate of cooling. And draw the parallel CF, and join AF intersecting BC in G; BG will denote the

the relative time in which the body will lose a certain part of its heat. Or describe the rectangular hyperbola CH, of which C is the centre, and AE the assymptote; then EH will mark that same relative time or range of cooling, for EH is obviously equal to BG. We thence gather, that even a moderate wind will quadruple the waste of heat, and that a vehement hurricane is capable of increasing the rate of dissipation perhaps fifteen or twenty times. Hence also the keen impression of frost winds on our feelings, and their prodigious effects in chilling the surface of the ground. We thus perceive, in a strong light, the vast utility of shelter, conspiring with the genial influence of the sun.

From the same principle we derive the construction of a new and very simple kind of *anemometer*. It is in reality nothing more than a thermometer, only with its bulb larger than usual. Holding it in the open still air, the temperature is marked: it is then warmed by the application of the hand, and the time is noted which it takes to sink back to the middle point. This I shall term the fundamental measure of cooling.

cooling. The same observation is made on exposing the bulb to the impression of the wind, and I shall call the time required for the bisection of the interval of temperatures, the occasional measure of cooling After these preliminaries, we have the following easy rule :—*Divide the fundamental by the occasional measure of cooling, and the excess of the quotient above unit, being multiplied by* 4½, *will express the velocity of the wind in miles* per *hour.* The bulb of the thermometer ought to be more than half an inch in diameter, and may, for the sake of portability, be filled with alcohol, tinged, as usual, with archil. To simplify the observation, a sliding scale of equal parts may be applied to the tube. When the bulb has acquired the due temperature, the zero of the slide is set opposite to the limit of the coloured liquor in the stem; and, after having been heated, it again stands at 20° in its descent, the time which it thence takes until it sinks to 10° is measured by a stop-watch. Extemporaneous calculation may be avoided, by having a table engraved upon the scale for the series of occasional intervals of cooling.

The

The principle which we have employed might be likewise extended to the case of running water. From some rough trials, I judge that the ordinary rate of cooling is doubled by the impression of a current which flows with the celerity of about half a foot *per* second, or one third of a mile *per* hour. The *formula* would, therefore, be $v = \frac{T-t}{3t}$. Thus very small velocities could be rendered sensible; but, as water consumes heat so much faster than air, it would require a large mass for immersion. Other precautions might be rendered necessary: I think it superfluous, however, to dwell any longer on a subject so minute.

CHAP.

## CHAPTER XV.

IT appears, then, that the loss of heat which a body sustains in its flight through the atmosphere is proportional to the extent of space which it describes, or the quantity of air which it displaces. This experimental result agrees with what we should expect from other considerations. But if, as the common theory supposes, the heated projectile were actually to touch in succession every portion of the whole fluid mass that lies in its track, we might expect something more than mere proportionality. Each molecule of air enveloped in the body's sweep, must, by suffering contact with the hot surface, acquire likewise its temperature. The measure of heat, therefore, which is at every instant transferred, ought to be exactly equal to what would be necessary to affect, in the same degree, the portion of fluid encountered during that interval. This inference deserves

deserves particular examination: I shall adopt, for a basis of comparison, the fact already noticed, namely, that a globe of four inches in diameter filled with warm water, and carried through the atmosphere with a velocity of 20 feet *per* second, loses, from the influence of this impulsion, at the rate of the fortieth part of its heat in a minute. A four inch sphere is equal to a cylinder of the same diameter, but whose altitude is $\frac{2}{3} \times 4$, or $\frac{8}{3}$ of an inch. Consequently the fortieth part of this, or the 15th of an inch, will be the thickness of a disc of water, which, at the same temperature, would contain a quantity of heat equal to what is every minute consumed. But, according to the common estimate, water contains 500 times more heat than atmospheric air of the same bulk and temperature. The portion of heat, therefore, which is transferred from the ball in the space of one minute, is equal to what would be sufficient to affect in the same degree a column of air whose altitude is $= 500 \times \frac{1}{15}$, or $33\frac{1}{3}$ inches. But the heated ball travels in a minute through an extent of 1200 feet, or 14400 inches; and

and consequently, since $\frac{14400}{33\frac{1}{3}} = 432$, it must communicate its impression to the 432dth part only of the interjacent air.

The numerical accuracy of this result will depend, it is evident, on the correctness of the ordinary estimate of the air's capacity for heat. From some observations, however, which I had occasion to make in the course of my hygrometrical researches, I am inclined to reckon that quality, or the measure of relative attraction, about three times less than is usually supposed. Yet after admitting this modification, it would still follow, that only the 144th part of the whole track, or column of air which the ball displaced in its flight, was really efficient in abstracting the heat.

We may therefore safely conclude, that, in the case of a ball carried swiftly through the atmosphere, not the hundredth part of the impinging fluid is fully exerted in cooling it. Nor can such a remarkable apparent deficiency be explained, by supposing each accession of air to require a certain definite portion of time to produce its effect; for the rapidity of the projectile's motion would then have no influence whatever in accelerating the dispersion of heat. If a greater number of aërial particles might, in a given time, strike against the hot surface, their contact would be proportionally of short duration, and consequently the measure of their action would be likewise only partial. But it seems far more reasonable to presume, that the successive molecules will exert their whole impression, or absorb heat to the point of saturation. Hence each, on arriving at the surface of the ball, would stop for some limited space, and therefore, during that interval however small, would bar all access of the subsequent parts of the current. This minute portion of time being expired, another contact would again ensue; and thus the progress would be continually repeated. The number of successive contacts, or the quantity of heat consumed, must consequently be proportional merely to the absolute time elapsed, independent altogether of the rapidity or slowness of the impinging fluid. But this legitimate inference is utterly inconsistent with observation. The momentary abstraction of heat was found in similar circumstances to be exactly proportioned to the velocity of impact.

pact. The instant of time in which each succeeding particle of air exerts its energy must, therefore, be smaller than the interval that corresponds to the swiftest motion experienced in our atmosphère. Thus are we compelled to admit, that the progressive flight of the ball accelerates the dispersion of its heat, only by multiplying, or more quickly renewing the contact, with fresh portions of the opposing fluid.

It is obvious that the conditions are not essentially altered, whether we suppose a current of air to strike against the ball at rest, or the ball itself to be transported with equal celerity through the still atmosphere. The relative motion, on which alone the effect must depend, is, in either case, precisely the same. And if we proceed to examine the matter with nice attention, instead of feeling surprise that the impinging fluid should exert an influence so partial, we shall find it more difficult to explain how it is enabled to make such a copious impression. The ordinary theory of resistance is not less defective in principle than discordant with observation. To treat this subject in the manner which its importance deserves would prove an attempt of the most arduous and extensive kind. But, for my present object, it will suffice to take a more general survey.

When a plane surface is exposed directly to the action of a stream of air, the particles, as they successively arrive, deposit their whole impulsive energy; and the resistance or pressure thence experienced, is supposed to consist merely in the force consumed in stemming the current. But though the fluid molecules may have their progressive motion extinguished, they are not therefore themselves annihilated. They will continue to accumulate on the solid obstacle, till the augmented elasticity arising from their mutual approximation generates a lateral or diverging efflux equivalent to the momentary accessions of the stream. A dense atmosphere is thus formed to a considerable depth above the obstructing surface; and the acquired pressure or repulsion of this incumbent stratum constitutes the real force of resistance. But the concentrated mass will be neither of uniform density, nor terminated by any precise boundary. Its condensation at right angles

angles to the surface, will be distinct from that which takes place in the parallel direction. In both lines, the intensity will vary by insensible shades, though after a different order and progression. This atmospheric accumulation will evidently be most protuberant about the centre, and will decline on either side. It must not only destroy the perpendicular appulse of the stream, but likewise impress a parallel motion; or rather, by exciting some modifying influence, it must convert the former into the latter. Nor will that effect be produced by a sudden deflection; the elasticity of the condensed portions of fluid being gradually exerted, will cause each individual streamlet to bend aside in a gentle curve. Each particle of air, following exactly the trace of that which precedes it, will have no tendency either to retard or accelerate its motion, but will suffer a lateral action calculated to divert its course by imperceptible degrees. But it is a beautiful and important property in dynamics, that the celerity of a point is not altered at all, when deflected by the gradual operation of perpendicular forces. From the composition of motion in the diagonal line, we learn, that, in every finite change of direction, the velocity is increased in the proportion of radius to the secant of the angle of deviation. But if that angle be continually diminished, its secant will approach rapidly to a ratio of equality with the radius, from which indeed it differs only by a quantity as the square of the angle. Thus, if we conceive a perpendicular streamlet to be diverted into a parallel course by the application of ten succeeding lateral impressions, its celerity after each deflection will be augmented in the proportion of the secant of 9°. Consequently the final celerity will be as (Sec. 9 ) , or 1.13884. And each of those angles being divided into ten equal parts, the celerity will, at every successive bend, be increased as the secant of 54′, and therefore will ultimately be as $(\text{Sec. } 54')^{100}$, or 1.012418. But if each of these were again subdivided into ten equal parts, the final increase of the celerity would be as $(\text{Sec. } 5' 24'')^{1000}$ = 1.001235, not amounting to the eight hundredth part of the whole. At each decimation,

tion, the excess is thus diminished more than tenfold, and consequently, by a repeated process, it would be totally extinguished.

It hence appears, that, by the divergent elasticity of the compressed atmospheric stratum, the direct appulse of the stream is gently changed into a parallel motion, of the same celerity. If the surface opposed to the current be a square, the depth of that stratum will, therefore, be equal to half its breadth. In other figures, the depth of the accumulated atmosphere will, after certain proportions, approximate to the semidiameter. This atmosphere, forming really a protuberant heap, will not be all of equal depth, nor will its lateral filaments shoot off in lines strictly parallel to the surface, but rather in hyperbolic curves, verging slowly towards that boundary as their assymptote. We need not, however, pursue such curious speculations. It is plain, that the terminating film of air which glides along the surface opposed to the stream can bear no sensible proportion to the whole accumulated mass or the contemporary deflected currents. Consequently, of the whole column of fluid whose impulsion is ex-

 pended,

pended, the portion that ever penetrates the limit of contact must, from this view, be extremely inconsiderable.

If the surface presents itself obliquely to the action of the stream, it will receive the shock of a narrower column; and therefore, the velocity of the lateral flow being still the same, the conglomerate atmosphere must have its depth proportionally smaller. Less force will likewise be required to produce the necessary deflection. The density of the protuberant fluid mass will hence be inferior, whether in the perpendicular or the parallel direction.

But the superincumbent strata, or the gliding films, which, deflected from their direct appulse, compose the current of atmospheric accumulation, will not continue to pursue their parallel motion, nor maintain unchanged their relative situations. The inmost particles, grazing along the opposing surface, will suffer such a constant train of impediments, as must quickly retard and consume their force; which being spent, they will be drawn back into the general stream, and their place occupied by others. Thus, in the shell of atmosphere

atmosphere which flows around the resisting body, there is excited likewise an internal circulation. The greater is the primary motion, the greater also must be the obstruction which the impinging particles will encounter. The partial interrupted slides, or the small spaces described by each of these upon the surface of contact, may, in every case, be the same; since any augmentation of velocity that can obtain, may be counterbalanced by the corresponding increased resistance.

This obstructing force is altogether independent of the quality or nature of the surface from which it originates. It is of the same kind as what takes place in the flow of water through extended tubes, or in the motion of elongated bodies through fluids. In practical hydraulics, it is well known that, without altering the column of pressure, the quantity of discharge is greatly diminished, by merely lengthening the conduit-pipe; and that a long cylinder is dragged through water with much more difficulty than a short one of equal diameter. Nor is the effect at all modified

by

by the peculiar properties of those cylinders, whether they are solid or hollow. Neither the adhesion of the fluid to its confining surface, nor the degree of smoothness or polish, seems to have any visible influence. An extended horizontal pipe, constructed of wood, will deliver as much water, as a similar one of lead: and though the experiment has not been tried, it cannot be doubted, that the motion of air through long narrow tubes would be found perfectly analogous to that of water. The flight of an arrow, shot through the atmosphere, is very sensibly impeded by the length of its shaft.

The sort of retardation which fluids experience in gliding over the surface of a solid obstacle is, therefore, distinct from resistance on the one hand, and from friction on the other, though more allied to the former. But clearly to trace its origin and mode of operation, will require a careful analysis of those several means wherewith Nature speedily extinguishes every motion upon earth, and seems to diffuse a principle of silence and repose; which made the ancients ascribe to matter

matter a sluggish inactivity, or rather an innate reluctance and inaptitude to change its place.* We shall perhaps find, that this prejudice, like many others, has some semblance of truth; and that even dead or inorganic substances must, in their recondite arrangements, exert such varying energies, and so like sensation itself, as, if fully unveiled to our eyes, could not fail to strike us with wonder and surprise. The resistance of fluids, or the force that is consumed in turning their particles aside from the course of the penetrating mass, we have already endeavoured to explain: we are now to investigate the cause and the conditions of friction, which obtains when one solid is drawn along the surface of another.

If the two surfaces which rub against each other are rough and uneven, there is a necessary waste of force, occasioned by the grinding and abrasion of their prominences. But friction subsists after the contiguous surfaces are worked down as regular and smooth as possible. In fact, the most elaborate polish can operate no other change than to diminish the size of the natural asperities. The

* See Note XXXI.

surface

surface of a body, being moulded by its internal structure, must evidently be furrowed, or toothed, or serrated. Friction is, therefore, commonly explained on the principle of the inclined plane, from the effort required to make the incumbent weight mount over a succession of eminences. But this explication, however currently repeated, is quite insufficient. The mass which is drawn along is not continually ascending; it must alternately rise and fall: for each superficial prominence will have a corresponding cavity; and since the boundary of contact is supposed to be horizontal, the total elevations will be equalled by their collateral depressions. Consequently, if the actuating force might suffer a perpetual diminution in lifting up the weight, it would, the next moment, receive an equal increase by letting it down again; and those opposite effects, destroying each other, could have no influence whatever on the general motion.

Adhesion seems still less capable of accounting for the origin of friction. A perpendicular force acting on a solid can evidently have no effect to impede its progress; and though this lateral force, owing

owing to the unavoidable inequalities of contact, may be subject to a certain irregular obliquity, the balance of chances must on the whole have the same tendency to accelerate, as to retard, the the motion. If the conterminous surfaces were, therefore, to remain absolutely passive, no friction could ever arise. Its existence demonstrates an unceasing mutual change of figure, the opposite planes, during the passage, continually seeking to accommodate themselves to all the minute and accidental varieties of contact. The one surface, being pressed against the other, becomes, as it were, compactly indented, by protruding some points and retracting others. This adaptation is not accomplished instantaneously, but requires very different periods to attain its *maximum*, according to the nature and relation of the substances concerned. In some cases, a few seconds are sufficient; in others, the full effect is not produced till after the lapse of several days. While the incumbent mass is drawn along, at every stage of its advance, it changes its external configuration, and approaches more or less towards a strict contiguity with the under surface. Hence the

the effort required to put it first in motion, and hence too the decreased measure of friction, which, if not deranged by adventitious causes, attends generally an augmented rapidity. This appears clearly established by the curious experiments of Coulomb, the most original and valuable which have been made on that interesting subject. Friction consists in the force expended to raise continually the surface of pressure by an oblique action. The upper surface travels over a perpetual system of inclined planes; but that system is ever changing, with alternate inversion. In this act, the incumbent weight makes incessant yet unavailing efforts to ascend: for the moment it has gained the summits of the superficial prominences, these sink down beneath it, and the adjoining cavities start up into elevations, presenting a new series of obstacles which are again to be surmounted; and thus the labours of Sisyphus are realized in the phænomena of friction.

The degree of friction must evidently depend on the angles of the natural protuberances, and which are determined by the elementary structure or the mutual relation of the two approxi-

mate

mate substances. The effect of polishing is only to abridge those asperities and increase their number, without altering in any respect their curvature or inflexions. The constant or successive acclivity produced by the ever-varying adaptation of the contiguous surfaces, remains, therefore, the same, and consequently the expense of force will still amount to the same proportion of the pressure. The intervention of a coat of oil, soap, or tallow, by readily accommodating itself to the variations of contact, must tend to equalize it, and therefore must lessen the angles, or soften the contour, of the successively emerging prominences, and thus diminish likewise the friction which thence results.

Such is apparently the real origin of friction. But the retardation which a fluid experiences in running over the surface of a solid, though derivable from the same source, is of a very different kind. The plane of mutual separation will in this case, too, be agitated by a similar alternating system of concatenated prominences and depressions, more extensive perhaps than in the former. From the want of cohesion among the fluid

fluid molecules, the transient and partial elevations at the bottom must be confined to their immediate vicinity; nor can the height or the pressure of the mass, exerting a balanced diffuse action, have any effect whatever to obstruct those minute displacements. The current, in gliding along, cannot maintain invariably the same relative disposition among its particles; the undermost stratum will at every stage assume a new arrangement, which must occasion an unavoidable waste of force. The law of this incessant expenditure is immediately deduced from the celebrated principle of the *conservatio virium vivarum:* it is always proportional to the square of the velocity of those secondary motions, or to the square of the velocity of the current itself; for the internal variations must evidently keep pace with the general motion, and acquire a corresponding share of its rapidity The same conclusion, however, may be derived from an elegant proposition in dynamics, to which we have already referred :—That, in all motions generated by insensible degrees, the increment of the square of the velocity is compounded of the ratio of the exciting

exciting force and the element of the space. But the internal derangements of the fluid stratum, arising from its proximity to the surface of termination, must in every case be similar, being formed on the scale of velocity by which the collective mass is urged along. And since, whatever such velocity may be, the minute spaces of aberration will evidently remain unaltered, the disturbing force, or the obstruction which the progressive motion must experience, is likewise as the square of that velocity.

This species of obstruction, therefore, results from the constitution of the fluid itself: it has no relation to the degree of pressure, but is determined merely by the extent of the surface of contact, and follows the duplicate ratio of the celerity of the general current. It often mingles its influence, however, with the distinct effects of ordinary friction. Thus, although the intervention of a coat of oil between two pieces of wood greatly reduces their measure of friction in the first instance, yet, as the rapidity of motion increases, the total resistance thence encountered becomes likewise very considerably augmented.

Nor is it required to suppose the unguent strictly fluid; the quality of softness, partaking of the nature of fluidity, will communicate in some degree analogous properties. With an application of soap or tallow, the apparent friction is found to increase in high velocities. But, to a certain extent, however limited, the same principle will obtain, even when the rubbing substances seem most remote from the character of fluidity. It was shown that the attrition of solids is caused and accompanied, by a perpetual series of alternating sympathetic motions between the two proximate surfaces. Those motions will in most cases be extremely minute, for the varying indentations have only to sustain a part of the incumbent weight, and the smallest impressions of the hard substance are able to exert an adequate counterbalancing repulsion. Friction, we have seen, consists in the constant effort to drag the loaded surface over a range of prominences which emerge in prolonged succession. But even where no pressure is applied, the mere sliding of conterminous planes against each other will be attended with some slight measure of impediment, arising

from

from the small portion of force which is consumed in generating the concomitant superficial vermiculations. These disturbing movements depend on the extent of surface and the degree of celerity. Hence it is that, in the majority of cases, the compounded friction, instead of rather diminishing with an increased velocity, continues the same, or perhaps acquires some augmentation. This is most perceptible where the pressure is comparatively small, the surface large, the motion rapid, and one or both of the proximate substances composed of soft materials.

Since, on a surface of given extent, the obstruction which a fluid experiences in gliding along, is proportional to the square of its velocity, and in that ratio likewise is the quantity of force expended in the shock of a fluid against a solid obstacle, the effect of attrition is exactly the same, as if certain portions of the mass had, at equal intervals of time, their progressive motion extinguished. The deranging forces, it is evident, are only exerted on the contiguous stratum: at every succeeding moment, therefore, a bundle of short filaments will spend their impetus, and retire into the general stream. Thus, whatever may be the rate of impulsion, each particle of fluid that reaches the margin of contact, will only travel over a certain limited space. The quickness of those reiterated applications will consequently be proportional to the velocity of the mass.

It hence appears, that each particle of the current must successively approach the bounding surface, and there, sliding to a certain minute distance, will spend its force, and again mingle in the body of the stream. The whole mass will, therefore, atchieve its contact in a space which is proportioned to the quantity of section. Comparing the experiments on the flow of water through conduit-pipes, as recited in Bossut's *Hydrodynamique*, I find, after making the proper reductions, that the velocity of projection from the bottom of a cistern is diminished about five times in the passage through an horizontal tube of one inch in diameter, and fifteen feet long. Consequently, while one part of the actuating force is discharged from the orifice, twenty-four parts are consumed in gliding against the sides of the pipe. Every particle contained must hence have repeated

peated its contact no less than twenty-four times before it made its escape; that is, the whole column of fluid must have inverted its internal arrangement at each interval of $7\frac{1}{2}$ inches. But the vertical section, and consequently the effect, would be the same, if the cylindrical rim were rolled flat and the water spread on it to the height of a quarter of an inch. Wherefore the distance to which each particle slides in succession, is thirty times the thickness of the film primarily affected. —The motion of air through long tubes or over extensive surfaces, no doubt, suffers a similar derangement; but I am not acquainted with any sufficient data to determine the relation of the thickness of the proximate film to the length of its successive slide.

The laborious investigation, now concluded, affords a clear and consistent explication of the mode by which a cold stream of air or water accelerates refrigeration. The whole turns on two capital points: 1. the several filaments of the current are gently diverted, and made to ply along the surface of the body with undiminished celerity, till they finally launch off and resume their flight from the farther side: and, 2, each portion of fluid that grazes against the obstacle, whatever might be its original force of impulsion, only sweeps a certain limited space, and then mingles in the general mass; during which contact, it must likewise abstract its share of heat, and, if it should come to touch again, it has in the interval dispersed its charge, and is fitted, therefore, to repeat the same impression. Hence the frequency of contact, and consequently the refrigerating power of the stream, is proportional to its appulsive velocity. If we conceive the surface of the body to be divided by a multitude of circumscribing lines, extremely near each other, yet equidistant; the total quantity of contact will be as the sum of these lines: but the whole extent of surface itself is obviously equal to the rectangle of that aggregate line, and the common breadth of such elementary zones. Thus, the influence of a current of fluid in cooling a body of any shape, however irregular, if not terminated by numerous

numerous and abrupt asperities, is proportional merely to the surface. The combined refrigerating action is, consequently, in the compound ratio of the surface and the velocity of impulse; and this theoretical deduction was found to be perfectly confirmed by observation.

## CHAPTER XVI.

WHEN a body is set to cool in a close room, the atmospheric shell which encases it, becoming heated and of course expanded, rises upwards, with a slow but regular motion. This gentle ascent of the column from beneath will, therefore, have the same influence as the impact of a current of air flowing at the same rate. If the body is colder than the external fluid, the proximate stratum, being chilled and contracted, will cause a tendency to descend; and the gradual accession of the stream will exert a similar effect as before, though in a reversed order. But these vertical motions are evidently very small, for the power of buoyancy is only distinguishable by a nice balance. They are altogether insufficient to explain that accelerated energy of refrigeration which was observed to take place in the higher temperatures, Such effect is derived from another source.—

Each

Each portion of air or gaseous fluid which touches a hot surface must receive that same measure of heat and a corresponding increase of elasticity. It, consequently, dilates with a force proportional to the space through which it recedes, or to the elevation of temperature which it has assumed. But the square of the acquired velocity, as we formerly remarked, is compounded of the space and the actuating force: in the present case, it is, therefore, as the square of either of these elements, or as the square of the degree of heat which is absorbed. The velocity of propulsion is hence proportional simply to the excess of temperature. The time of action is always evidently the same, because, if the space be enlarged, the rate of dilatation is likewise increased; and hence, from every exciting point of the hot surface, a slender continued stream of air is emitted perpendicularly, whose velocity is proportioned to the measure of heat incessantly communicated. When the process is inverted, and the surface affected is colder than the surrounding atmosphere, the contiguous portions suffer contraction and a diminution

diminution of their elasticity, which occasions a gentle perpendicular flow directed towards its source, and productive of a similar though an opposite effect.

Thus the discharge of heat from a body is materially promoted by the soft propellent motion excited continually at its surface. This efflux extends to a very short distance, before it spends its force and loses itself in the atmosphere; yet it equally produces the refrigerating effect, by quickening the circulation and fresh contact of the ambient medium. Though it conspires with pulsation to accelerate the dispersion of heat, it differs essentially in its character from that species of energy. Pulsation is the same at all degrees of heat, and its intensity depends merely on the nature of the bounding surface: but the perpendicular flow is more vigorous in proportion to the excess of temperature, and has no relation whatever to the qualities, physical or mechanical, of that surface. It was shown that, only a very few particles disseminated in the contiguous shell of air, feel at once the pulsatory influence: the other

other particles, which constitute the general mass, probably imbibe their share of heat, and passively obey the impression of their augmented elasticity.

From a comparison of numerous trials made with canisters of various shapes and dimensions, and filled with boiling water, I find, at the equidistant temperatures of 10, 40, and 70 degrees, reckoning from the standard of the external air, that, with a metallic surface, the rates of cooling are very nearly as 2, 3, and 4; but, when the surface is papered or covered with a coat of lamp-black, the rates of cooling are respectively as the numbers 4, 5, and 6. Thus, in either case, the gentle perpendicular flow of heated air, corresponding to an excess of 30° of temperature, has an action as 1; and the double of this, with a similar excess of 60: it, therefore, exerts effects which are exactly proportioned to its expansion or augmented elasticity. The energy that a surface of paper communicates by exciting copious pulsations, is constantly denoted by 2; yet its influence is comparatively small in the higher temperatures. With an excess of 10° above the standard of the surrounding air, a papered surface cools

cools twice as fast as one of planished tin; but, with an excess of 70°, it cools only one half faster.

It was formerly shown, that the pulsatory action of paper is eight times greater than that of tin; consequently this species of energy exerted by the metallic surface is expressed by $\frac{2}{7}$, which, being subtracted from 2, gives $1\frac{5}{7}$ for the refrigerating power of such a surface with an excess of 10° of temperature, independent of the auxiliary effect of pulsation. But the flow of air corresponding to that excess must have an influence in abstracting heat, which is denoted by $\frac{1}{3}$; and hence, after deducting those external impressions, whether produced by pulsation or actual motion, there still remains $1\frac{8}{21}$, for the power apparently inherent with which every substance tends to an equilibrium of temperature.—A similar conclusion is derived in a manner somewhat different. I selected a mercurial thermometer with a large bulb and a slender stem, to which was adapted a moveable scale containing only a few degrees, but these nearly each half an inch in length. Having gilt the bulb with silver leaf, and set the zero of the

the scale to correspond with the temperature of the room, I applied the heat of my hand, and then, leaving the instrument to cool, I observed its progress with a stop-watch. From 8° to 4°, it took 135 seconds; thence to 2°, it required 142; and the last bisection, till it reached 1°, was performed in 146 seconds. Whence, at each observation, during which the heat is reduced to one half, the time elapsed is lengthened out, yet evidently with a retarding progression. In the first interval, there is an addition of seven seconds; in the next one, another increase of four seconds; and, if the process of bisection had been continually repeated, it seems probable that the successive augmentation would have advanced by halves. But it is the nature of such a progression that any term equals the sum of all which succeed to it. Consequently, 150 seconds is the ultimate limit to which the intervals of bisection approach.—The same experiment being repeated after the gilding had been rubbed off, the time of cooling from 8° to 4° was 65″, that from 4° to 2° was 67″, and that from 2° to 1° was 68″.

In

In the case of the naked ball, therefore, the ultimate limit would be about 69 seconds. And since $\frac{1}{128} = \frac{1}{150} - \frac{1}{69}$, the superior pulsatory effect of the vitreous surface amounts to the 128th part for each second. Consequently the single action of the metallic surface, in the same space of time, is only about the 900dth part, for $\frac{1}{7} \times \frac{1}{128} = \frac{1}{896}$. But, $\frac{1}{150} - \frac{1}{900} = \frac{1}{180}$; and hence, besides the heat abstracted from the ball by the pulsatory and regressive motions of the surrounding air, there is some other mode by which it is dispersed at the rate of the 180th part each second.

The portion of heat thus consumed is most certainly not annihilated; neither is it transported to a distance, by any species of elastic motion excited in the encircling fluid. It is, therefore, absorbed by the contiguous shell of matter, and afterwards slowly diffused through the extended mass. Air is still the sole medium by which heat endeavours to maintain the balance among remote or detached bodies; but here its operation is of a passive nature, and it receives and

conveys

conveys the calorific impressions through its substance in the same manner as a bar of iron or any solid material.

This completes the analysis of the refrigerating action of air. There are four distinct modes in which it produces the effect: three of these are always conjoined, and the fourth only throws in its occasional influence. They all conspire to the same end, but their relative shares of operation are various and mutable. One source of communication depends on the quality of the heated surface, another on its elevation of temperature, a third on the permanent conducting disposition of the air, and the last arises from the celerity of impulse by which that active fluid may chance to be affected. The continual ascent of the hot, and consequently rarefied, air, must contribute in some degree, though indirectly, to accelerate the effect; for it is evident, that the stagnation of a warm encircling atmosphere would debilitate the operation of the combined refrigerating causes.

Having developed the separate influence of those several distinct yet associated operations,

we

we are enabled now to determine their joint effect. The power which the air exerts in cooling a surface by the agency of its internal vibrations, and that which results from the ordinary process of the conducting of heat through the surrounding mass, are both of them constant, and may be included in the same estimate. But the refrigerating energy derived from the slow expansive recession of the heated molecules of air is continually decreasing, being proportioned simply to the excess of temperature. This expenditure of heat occasioned by the varying reiteration of aërial contact, is always inseparably conjoined, however, with that which is produced by the pulsation and diffusive absorption of the atmospheric mass. Their combined action will, therefore, be expressed by a constant quantity annexed to the declining measure of temperature. It was already noticed that the refrigerating power of the perpendicular flow of air, which corresponds to 30 degrees of heat, being as 1, the aggregate effect of the joint cooling processes at 10° centigrade is, on a metallic surface, as 2, and, on a surface of paper, as 4. Consequently, with that excess,

excess, the whole refrigerating energy exerted by still air on a metallic surface will be denoted by 60, and on a surface of paper, by 120: at the limit of equilibrium, therefore, the respective energies, being each diminished by 10, will be expressed by the numbers 50 and 110. Hence the rates of cooling that correspond in general to any height, *h*, of temperature are, for a surface of metal and one of paper, represented by $50 + h$ and $110 + h$.

It is obvious that, near the limit of equilibrium, *h* is comparatively small, and may be rejected without sensible error. The influence of the superficial shifting of air is, in this case, almost extinguished; the residual measure of heat forms, at equal intervals of time, very nearly a descending geometrical progression; and, under similar circumstances, the rate of cooling of a surface of tin is to that of a surface of paper, as 5 is to 11, or is rather less than the half. In very high temperatures, on the contrary, the constant numbers annexed to *h* will have relatively but small effect. The pulsatory transfer of heat will now be lost in comparison with the other accelerated sources of its discharge, and the rates of cooling with different surfaces will hence approach to a ratio of equality. Thus, the surrounding air, being supposed at the point of congelation, a tin ball filled with boiling water, if clear, will cool at the rate of 150, and if painted, at the rate of 210; or after the proportion of 5 to 7: but, when filled with boiling oil, the rates of cooling will be respectively as 300 + 50 and 300 + 110, or will bear the ratio of 6 to 7 very nearly. In those elevated temperatures, the progress of cooling will follow a singular law. The refrigerating energy being nearly as $h$, and its intensity of impression being likewise as $h$, the decrement of heat must evidently be compounded of both these and the element of the time. Therefore, $-dh = h^2dt$, and $\frac{-dh}{h^2} = dt$; of which the integral is $\frac{1}{h} = t$, and the complete integral $t = \frac{1}{h} - \frac{1}{H}$, where $t$ denotes the time elapsed between any two high temperatures H and $h$. We hence derive this paradoxical conclusion, that, from whatever degree of heat a body begins its descent, it will reach the same point of temperature in some finite

finite time. In fact, the limit of the expression $\frac{1}{h} - \frac{1}{H}$ is invariably the fraction $\frac{1}{h}$, although H should transcend the bounds of numeration. The passage through temperatures of extreme elevation is performed with the utmost rapidity. Thus, between 10,000° and 1,000° the time spent is ten times less than between that stage and 100°; the former being represented by $\frac{9}{10,000}$, and the latter by $\frac{9}{1,000}$. This inference is perfectly consonant with observation. To whatever degree of heat a charged crucible has been pushed, it will, after being withdrawn, take very nearly the same time to sink to some fixed point, such as that of boiling water. It was the late Mr. Wedgwood's elegant and valuable invention of the pyrometer that first opened to our view the immense range of the rising scale of heat, and disclosed the vast extent of the power of chemical furnaces.

But it is easy to discover the general relation which connects the time and temperature. The refrigerating energy is always expressed by $a + h$, $a$ denoting the constant additive number, which is 50 for a surface of metal, and 110 for one of paper. The intensity of impression is evidently as $h$, the excess of temperature itself: wherefore, $-dh = (a + h)\, h.dt$, and $dt = \frac{-dh}{ah + h^2}$, which, being integrated, gives $t = \frac{1}{a}\left(\text{Log. } \frac{h}{a+h}\right)$, and the complete integral is consequently $t = \frac{1}{a} \left(\text{Log. } \frac{H}{h} - \text{Log. } \frac{a+H}{a+h}\right)$* This formula is abundantly simple. To find the time which any body immersed in still air takes to cool,—*From the difference of the logarithms of the initial and final temperatures, counting from that of the encircling fluid, subtract the difference of the logarithms of those temperatures augmented each by a constant number, and the remainder, being divided by the same number, will give a quotient proportional to the time elapsed.* This number is 50, in the case of a surface of tin, and 110, in that of a surface of paper; in other intermediate instances, it will incline to the former or to the latter, according as the exterior coat partakes more of the metallic nature, or approaches rather to the condition of earths or vegetables. All the metals, I presume, are included within the

* See Note XXXII.

limits

limits of 45 and 55, and the great majority of other substances will not rank below 100. For a vitreous surface, the constant additive number may be 105. The augments above stated will hence nearly comprehend every possible case, and the computation for them happens to be singularly direct and expeditious. If from the difference of the logarithms of the initial and final temperatures, we subtract the difference of the logarithms of those temperatures increased each by 50, and reckon as integers the first three figures after the decimal point; the remainder will express in minutes the time required in cooling by a hollow tin ball filled with hot water, and of six inches diameter. And if 110 be annexed to the limiting temperatures, and the result multiplied by 5 and divided by 11, we shall obtain the time which the same ball would take to cool, after having its surface coated with lamp-black. In this way, I have constructed a table extending 100 degrees, or to the interval between freezing and boiling water.

TABLE of the progressive cooling of a hollow tIn ball, of six inches in diameter, and filled with boiling water, whether the surface is clear, or covered with a coat of pigment.

| Heat. | Metal. | Paper. | Heat. | Metal. | Paper. | Heat. | Metal. | Paper. | Heat. | Metal. | Paper. |
|---|---|---|---|---|---|---|---|---|---|---|---|
| 100 | | | 75 | 45.8 | 31.8 | 50 | 124.9 | 83.2 | 25 | 301.0 | 186.4 |
| 99 | 1.5 | 1.0 | 74 | 48.1 | 33.4 | 49 | 129.3 | 85.9 | 24 | 312.9 | 193.0 |
| 98 | 3.0 | 2.1 | 73 | 50.5 | 35.0 | 48 | 133.9 | 88.7 | 23 | 325.6 | 200.0 |
| 97 | 4.5 | 3.2 | 72 | 52.9 | 36.6 | 47 | 138.6 | 91.6 | 22 | 338.8 | 207.2 |
| 96 | 6.0 | 4.3 | 71 | 55.4 | 38.3 | 46 | 143.4 | 94.6 | 21 | 353.0 | 214.9 |
| 95 | 7.6 | 5.4 | 70 | 58.0 | 40.0 | 45 | 148.4 | 97.7 | 20 | 368.0 | 223.0 |
| 94 | 9.2 | 6.5 | 69 | 60.6 | 41.7 | 44 | 153.6 | 100.8 | 19 | 384.0 | 231.6 |
| 93 | 10.8 | 7.6 | 68 | 63.3 | 43.5 | 43 | 158.9 | 104.1 | 18 | 401.1 | 240.8 |
| 92 | 12.1 | 8.8 | 67 | 66.0 | 45.3 | 42 | 164.5 | 107.5 | 17 | 419.5 | 250.5 |
| 91 | 14.1 | 10.0 | 66 | 68.8 | 47.2 | 41 | 170.2 | 110.9 | 16 | 439.3 | 260.9 |
| 90 | 15.8 | 11.2 | 65 | 71.7 | 49.1 | 40 | 176.1 | 114.5 | 15 | 460.7 | 272.1 |
| 89 | 17.5 | 12.4 | 64 | 74.6 | 51.0 | 39 | 182.2 | 118.1 | 14 | 484.0 | 284.1 |
| 88 | 19.3 | 13.6 | 63 | 77.6 | 53.0 | 38 | 188.6 | 121.9 | 13 | 509.3 | 297.2 |
| 87 | 21.1 | 14.9 | 62 | 80.7 | 55.0 | 37 | 195.2 | 125.8 | 12 | 537.1 | 311.4 |
| 86 | 22.9 | 16.2 | 61 | 83.9 | 57.0 | 36 | 202.1 | 129.9 | 11 | 567.8 | 326.9 |
| 85 | 24.8 | 17.5 | 60 | 87.2 | 59.1 | 35 | 209.3 | 134.1 | 10 | 602.0 | 344.1 |
| 84 | 26.7 | 18.8 | 59 | 90.5 | 61.3 | 34 | 216.7 | 138.4 | 9 | 640.5 | 363.2 |
| 83 | 28.7 | 20.1 | 58 | 93.9 | 63.5 | 33 | 224.5 | 143.0 | 8 | 684.3 | 384.8 |
| 82 | 30.7 | 21.5 | 57 | 97.4 | 65.7 | 32 | 232.6 | 147.7 | 7 | 734.7 | 409.5 |
| 81 | 32.7 | 22.9 | 56 | 101.0 | 68.0 | 31 | 241.0 | 152.6 | 6 | 794.0 | 438.2 |
| 80 | 34.8 | 24.3 | 55 | 104.7 | 70.4 | 30 | 250.0 | 157.6 | 5 | 865.3 | 472.5 |
| 79 | 36.9 | 25.7 | 54 | 108.5 | 72.8 | 29 | 259.1 | 162.9 | 4 | 954.2 | 514.8 |
| 78 | 39.0 | 27.2 | 53 | 112.5 | 75.3 | 28 | 268.8 | 168.4 | 3 | 1071.1 | 570.3 |
| 77 | 41.2 | 28.7 | 52 | 116.5 | 77.9 | 27 | 279.0 | 174.2 | 2 | 1238.9 | 648.2 |
| 76 | 43.5 | 30.2 | 51 | 120.7 | 80.5 | 26 | 289.7 | 180.2 | 1 | 1531.5 | 783.2 |

This

This table is easily adapted to vessels of any shape or dimensions. If the surface of a six inch sphere were rolled out into a plane, it would cover a circle of a foot in diameter; and if the matter contained were spread equally over that extent, it would form a cylinder of an inch in depth. The measures, whether superficial or solid, of such a sphere, are thus expressed in inches by the same numbers; and hence, to find the time of cooling that belongs to a close vessel of any form or size, we have only to multiply the corresponding differences in the table by the capacity and divide by the surface. I shall illustrate this by an example.—Suppose a cylindrical tin vessel, eight inches wide and twelve inches high, and filled with hot water, to be placed in a room of the temperature of 18°, and it were required to determine the time which it would take to cool from 64 to 30 degrees. The excesses of temperature are here 46° and 12°, opposite to which, in the column of metal, are 143′.4 and 537′.1; the difference is 393′.7, which denotes the interval of time due to a sphere of bright tin six inches in diameter. And, in the column of paper, the corresponding quantities are 94.6 and 311′.4, whose difference 216′.8 will express the time due to the same sphere when painted. But the surface of the cylindrical vessel is measured by $128 \times \frac{\pi}{4} + 8\pi \times 12 = 120\pi$, and its capacity by $64 \times 12 \times \frac{\pi}{4} = 192\pi$; and since $\frac{192\pi}{120\pi} = 1.6$, if the former differences be multiplied by this fractional number, we shall obtain the quantities sought, and which are, for those two kinds of surfaces, respectively 630 and 347.

We are now able to compute the separate values oı effects of those elementary energies which are concerned in the propagation of heat. The general *formula* for a tin surface is $t = 1000$ M $(\text{Log.} \frac{h}{50 + h})$, whose differential is $dt = 1000$ M $(\frac{dh}{h} - \frac{dh}{50 + h})$, or $dt = \frac{50}{50 + h} (\frac{dh}{h})$ 1000 M; and, for a surface of paper, the differential becomes $dt = \frac{5}{11} \times \frac{110}{110 + h} \times 1000$ M, or $dt = \frac{50}{110 + h} \times 1000$ M. But we have this analogy, as the decrement of the heat is to the decrement of the time, so is the whole heat, to the *range*, or the time in which an equilibrium of temperature would

would be attained if the process were continued uniformly with its first intensity. Consequently, for a six inch ball of clear tin filled with boiling water, the portion of heat spent every minute is $\frac{50+h}{50}\left(\frac{1}{1000\text{ M}}\right)$, or at the rate of the $\frac{50+h}{50}$ $\left(\frac{1}{434.2945}\right)$ part; and when the surface is covered with paper, the rate of cooling is $\frac{110+h}{50}$ $\left(\frac{1}{434.2945}\right)$ part *per* minute. Of both these, the variable portion is $\frac{h}{50}\left(\frac{1}{434.2945}\right)$, which measures the effect of the slow perpendicular motion excited in the contiguous atmosphere. Deducting this from each, it appears, that the expenditure from a metallic surface towards the limit of equilibrium is equal to $\frac{1}{434.2945}$, and that from a painted one is $\frac{110}{50}\left(\frac{1}{434.2945}\right)$; whence the former in every case exceeds the latter by $\frac{60}{50}$ $\left(\frac{1}{434.2945}\right)$, or the 361.912th part. This difference proceeds merely from the unequal energy of pulsation; and, therefore, the pulsatory discharge every minute from the metallic surface is is $\frac{60}{350}\left(\frac{1}{434.2945}\right)$, or the 2533.385th part; but, from the painted surface, it is $\frac{480}{350}\left(\frac{1}{434.2945}\right)$, or the 316.673dth part of the whole residual heat. If the former of these be subtracted from the effect of a tin surface at the point of quiescence, we shall obtain $\frac{290}{350}\left(\frac{1}{434.2945}\right)$, or the 524.1485th part, for the measure of heat conducted away through the stationary mass of the surrounding air.—But it may be convenient to exhibit, in a collected view, those several results. From a hollow sphere, six inches in diameter, and filled with boiling water, the portions of heat discharged every minute, are thus represented:

By abduction, the 524.1485th;
By recession, the $h \times$ 21714.725th;
And by pulsation, the 2533.385th, from a metallic surface; and the 316.673th, from a surface of paper.

It hence appears, that the expenditure by abduction, or internal communication, is equivalent to what would be produced by a reflux of air charged

charged to the temperature of the surface, and moving with a celerity somewhat less than an inch *per* minute. for a column of air standing on a base equal to the surface of ball, and having 500 inches in altitude, would contain the same quantity of heat; and $\frac{500}{524.1485} = .9539$, or the $\frac{21}{22}$dth part of an inch very nearly. The velocity of recession, or the slow perpendicular motion excited in the elastic fluid is, for each degree of excess of temperature, at the rate of an inch in the space of $\frac{21714.725}{500}$ minutes, or 43.43; and hence, corresponding to the interval between boiling and freezing, the celerity of flow is 2.3 inches every minute. The effect of pulsation is estimated with equal facility; for $\frac{2533.385}{500} = 5\frac{1}{15}$, or the pulsatory energy from a surface of tin is equivalent to the abstraction of a continued flow of air, having its full charge of heat, and the velocity of an inch in five minutes.

The various disposition of different surfaces to excite certain tremulous impressions in the atmosphere is, therefore, the source of whatever diversity of power that appears in shedding their heat.

But

But those vibratory dispersive energies must evidently have the greatest comparative effect, when the auxiliary repellent flow of heated air becomes languid; that is, in the low temperatures, or near the limit of quiescence. A coating of tin, and one of paper, may exhibit the extreme rates of cooling; and other surfaces will occupy some intermediate stations, and will, according to their nature and condition, incline more to the one or the other. The relative position of different subtances, in that respect, is consequently determined by their peculiar physical qualities, modified, however, by the thickness or tenuity of their superficial stratum, and by the smoothness or striated outlines of their exterior boundary. I have not thought it necessary to pursue such comparison in detail; but the few observations which I shall select are sufficient to corroborate the general theory.

## EXPERIMENT XLVIII.

A bright tin canister, three inches square, and filled with boiling water, was set to cool in a close room:

room: in the space of 81 minutes, it sunk from 60 to 30 degrees above the temperature of the apartment. But when the sides were rubbed with mercury, and exhibited a resplendent lustre, it made the same descent in 78 minutes; and after the interval of a day or two, the tin, having, by repeated aspersions of mercury, been thoroughly penetrated by that fluid metal, and presenting a surface of a matt white, the process of cooling was performed in 72 minutes.

A mercurial surface has thus a decided influence in accelerating the progress of refrigeration. But it was formerly stated, that mercury surpasses all the metals by its energy of exciting aërial pulsations.—With respect to substances which are not metallic, they must approach very nearly to paper in their affections to heat. Glass itself seems not to differ by the twentieth part from that standard.

## EXPERIMENT XLIX.

Another tin canister, three inches square, and filled with boiling water, was set to cool as before. It

It took 105 minutes to sink from 30 to 15 degrees, above the temperature of the room. But, all the sides being carefully covered with bibulous paper soaked in olive oil, the same descent was made in 58 minutes. On removing these coatings, and covering the surface with a mere film of oil by means of a feather dipt in it, the cooling was not accomplished till after the lapse of 87 minutes.

The fact now recited evinces clearly the diminished effect resulting from the tenuity of the superficial film. The accelerating influence introduced by the thin layer of oil, which adheres to the metallic surface, is only the fourth part of what belongs to such a coating of the proper thickness. This singular modification is farther elucidated by another observation of a distinct yet kindred nature.

## EXPERIMENT L.

The same tin, which, in the space of 105 minutes, cooled down from an excess of 30 to that of of 15 degrees of temperature, was rubbed in one direction with fine sand-paper; it now made the same descent in 100 minutes. But, on being rubbed hard with a coarser sand-paper, the effect was performed in 96 minutes.

It was shown that, in consequence of a closer though partial contact with the bounding atmosphere, a striated surface of metal is fitted to excite more energetic pulsations. Hence the reason why those numerous flutings promote, to a certain degree, the operation of cooling. It is certainly not owing to an artificial increase of surface, for a glass vessel, treated in the same way, betrays no alteration in its rate of cooling.

But the modifying influence produced by the proximity of a metallic substratum, is most unequivocally displayed in the successive application of differently attenuated films of isinglass. I have made a series of observations directed to that object, and shall here present the results in one collected view.

## EXPERIMENT LI.

A canister of planished tin, three inches square, was filled with hot water, and placed upon a slender insulated stand, in a close room: it took 117 minutes to cool down from 20 to 10 degrees above the temperature of the apartment. The sides of the canister were then coated by repeated additions of dissolved isinglass, in the manner formerly described; and after this had dried into a thin pellicle, the corresponding rate of cooling was, at each gradation, carefully ascertained. The numerical progression was as follows:—

| Thickness of the pellicle, in parts of an inch. | Time of cooling from 20° to 10°, in minutes. |
|---|---|
| 50,000th | 101 |
| 20,000th | 89 |
| 10,000th | 80 |
| 5,000th | 72 |
| 2,000th | 66 |
| 1,000th | 63 |
| 500th | 62 |
| 300th, or more | 61 |

But

But these several results will be found, on examination, to accommodate themselves with surprising accuracy to the deductions of theory. At the limit of equilibrium, the refrigerating energy of a surface of tin, and that of one of paper, it was shown, are respectively as 50 and 110. For any other surface, therefore, this energy will be denoted by $110 - 60r$, if $r$ represents its reflecting power compared with that of tin. But the relative measure of reflection due to any given tenuity of pellicle was already determined by interpolation, and the mutual comparison of numerous examples. Thus, the reflective power of a coat of isinglass, the 10,000th of an inch thick, is .55, and $110 - 60r = 77$: the time of cooling ought, therefore, to be, for a globe of six inches in diameter, $= \frac{50}{77}$ (Log. $\frac{80}{10}$ — Log. $\frac{97}{87}$) $= 164\frac{1}{2}$, and the half of this, or 82′, is what corresponds to a canister of three inches. In the remaining cases, the coincidence appears equally striking. For example, when the thickness of the pellicle is the 2000th of an inch, $r$ is .17, and $110 - 60r = 100$; the time elapsed is, therefore, $= \frac{50}{100}$ (Log. $\frac{20}{10}$ — Log. $\frac{120}{110}$) $\frac{1}{2} = 66$, which agrees with the experimental result.*

* See Note XXXIII.

We can now explain an apparent anomaly which occurred near the commencement of our inquiry. It was established as a fundamental proposition, that every surface has an equal power both to absorb and to discharge heat. But while, the planished side of the canister fronting the reflector, the impression made upon the focal ball was only 12 degrees,—on turning the painted side to face the reflector, and covering the ball with tin-foil, the greatest effect amounted to 22 degrees. The procedure being here merely inverted, we might expect like results. The augmented action was evidently owing to this cause,—that a metallic surface cools much slower than one of glass, and is therefore proportionally more affected by the same energy of impression. If the focal ball had been covered with leaf silver, which forms a smooth and brilliant surface, the effect would have risen perhaps to $\frac{11}{5} \times 12$, or about 26. But tinfoil is generally in a slight degree oxydated, and, in applying itself to such a small ball, is gathered up into numerous folds: consequently its disposition to cool is, on both these accounts, something increased. From a variety of estimates, I reckon the rate of cooling with a sur-

face

face of tin-foil and one of glass as five to nine; but $\frac{9}{5} \times 12 = 21\frac{1}{5}$, which perfectly agrees with observation

We set out with assuming it as an undoubted principle, that the impression made upon the differential thermometer is proportional to the exciting force, or the momentary accessions of heat. This, however, requires some material limitations; nor is it strictly true, even with the same subject, if the range be considerable. And such is the usual progress of discovery: as new lights successively burst in upon us, we learn by degrees to correct our primary notions. The rate with which a body cools appears sensibly accelerated in the higher temperatures, especially in the case of a metallic surface. Consequently, the calorific action which comes to be counterbalanced by that refrigerating disposition, must always somewhat exceed the elevation of the liquor in the differential thermometer. Thus, when the effect upon the naked focal ball indicates 60, or 6° centigrade, the real measure of calorific energy is $\frac{111}{105} \times 60$, or $63\frac{1}{2}$ nearly. In general, the additive correction will be denoted by $\frac{x^2}{1050}$. Therefore,

fore, under 33, the difference will not amount to one division. But since the quantities obtained by actual experiment were very seldom found to exceed that number, the correction may in most instances be disregarded. As a theoretical inference, I have thought proper to notice it; yet, for that reason, to revise our prior deductions, and introduce such slight modifications, might well seem a fastidious refinement.—

I will not, however, quit this part of the subject without remarking, that the observation of the various celerity with which different bodies cool, affords incomparably the simplest, the most commodious, and, perhaps after all, the most accurate, method, of ascertaining their comparative attractions with respect to heat. For this purpose, a glass vessel must be selected, as thin as possible, of a moderate size and globular shape, to contain the substance to be submitted to examination. It should have a very short and narrow neck, adapted to receive a fine thermometer with a tapering or cylindrical bulb. This flask, being filled with hot water, and its thermometer inserted,

inserted, is placed on a slender insulated stand, under a large bell glass, and the time carefully noted of its passage between two known and pretty distant points of temperature, estimating always from that of the room. Filled again with some other liquid substance, the same observation is repeated. If less time be now required, the substance must contain a proportionally smaller share of heat. Reckoning water, therefore, as a standard, the relative portion of heat which impregnates an equal bulk of any other species of liquid may be thence easily deduced. To determine the specific heat of a solid substance, it must be granulated, or broken down to a gross powder, and introduced into the flask, the interstices being filled up with water. The interval elapsed between two capital divisions is observed as before; and the bulk of the included water being known, the corresponding time of cooling being subtracted, will give what exclusively belongs to the solid matter, which being then augmented in the proportion of the whole contents, will express the relative capacity of the substance thus examined. These results, however, exhibit only the quantities of heat contained in equal bulks; but, if they be divided by the specific gravities, the quotients will express the respective shares inherent in equal weights, or in equal portions of differently constituted matter. I might suggest a variety of precautions which would improve the accuracy of the procedure: for instance, the measure of heat contained in the substance of the flask itself, may be calculated and regularly deducted, and the observation of the relative progress of cooling may be repeated at certain capital divisions of the thermometer, and the mean of them all assumed. But I content myself with giving the spirit of the method, without entering on the practical details. From several trials which I have made, though upon a small scale, I have reason to be perfectly satisfied with the precision and facility of this plan of proceeding.*

* See Note XXXIV.

## CHAPTER XVII.

OF the three elementary powers which concur to carry forward the process of refrigeration in the medium of the atmosphere, the one which depends on the quality of surface, or the energy of pulsation, is entirely precluded in the case of a hot body immersed in water or other liquids. The operation of cooling is here performed by the combined action of the two remaining principles of dispersion. A portion of the heat is uniformly absorbed by the surrounding water, and conducted away through the internal mass, in the same manner as if this were congealed into solid ice. The remaining portion is discharged by the slow recession of the heated particles, or the perpendicular motion produced by their mutual distension. This latter force of consumption will evidently be variable, being proportioned to the degree of expansion, and therefore increasing with the rise of temperature.

Experiment establishes clearly the conjunction of these two distinct and heterogeneous modes of operation. The rapidity, however, with which bodies lose or acquire heat, when plunged in a bath of water, renders it extremely difficult to mark their progress with sufficient accuracy. Nor will I pretend to the same nicety as in the case of an atmospheric medium; but I may yet hope to obtain such a degree of approximation as will fully serve all the purposes of general investigation. I shall produce, as the basis of analytical deduction, the mean interpolated results of various observations.

## EXPERIMENT LI.

A hollow tin ball, four inches in diameter, with a narrow neck, being filled with boiling water, and having a long delicate thermometer, with a tapering bulb, inserted beyond the centre of the liquid, was plunged to the depth of eight or ten inches in a large tub of water at the point of congelation, and suffered to rest on a slender tin stool. In $2\frac{1}{2}$ minutes, the thermometer sunk from

from 70° to the mid division, or 35°; it took four minutes to drop from 40° to 20°, but $6\frac{1}{2}$ minutes were required to perform the final bisection from 10° to 5 .—

It is obvious, therefore, that the accelerated progression which thus obtains in the higher temperatures, must arise from the co-operation of some cause whose intensity is augmented with the degree of heat. This conspiring energy seems even to increase in a faster ratio than the excess of temperature; for, while these temperatures form an arithmetical series, the measure of the compound action, reckoning upwards, is denoted respectively by the fractions $\frac{1}{6\frac{1}{2}}$, $\frac{1}{4}$, and $\frac{1}{2\frac{1}{2}}$, and their differences constitute the fractions $\frac{5}{52}$ and $\frac{1}{20}$, or $\frac{1}{10}$ and $\frac{1}{6\frac{1}{2}}$ nearly, in which the rapid augmentation of power is most perceptible. But a similar conclusion, equally decisive, is derived from another and more striking method of observation.

EXPERI-

## EXPERIMENT LII.

The water bath was heated up successively to 30 and 60 degrees of temperature, and the ball with its inserted thermometer immersed as before. When the bath stood at zero, the thermometer required six minutes to drop from 20° to 10°; when it was raised to 30°, the surplus heat of the ball sunk from 50° to 40°, in 3¼ minutes; but when the water of the tub had attained the temperature of 60°, an equal effect, or a descent from 80° to 70°, was performed in the space of two minutes.—

In the medium of common air, those equal effects would have been produced in the same interval of time. A similar result is obtained, if we employ even a bath of oil or alcohol. The increased flow which, with the same difference of temperature, takes place in a mass of warm water, is an evident consequence of a property almost peculiar to that fluid, which, on receiving equal

equal accessions of heat, does not expand uniformly, but with a rapid acceleration. In reality, the dilatations of water corresponding to equal increments of heat, form very nearly an arithmetical progression; and therefore, reckoning from the origin of the scale, the whole measures of expansion must constitute a series of square numbers. It was shown already, that the celerity of the perpendicular flow is proportioned to the repellent force, or the distension required, and consequently in the case of the hot water bath, where the difference is only trifling between it and the immersed ball, the auxiliary action is proportioned likewise to the degree of heat. And this inference agrees with observation: the temperatures of 30° and 60° rise by equal ascents, but the corresponding effects acquire likewise equal augments nearly, for the difference between the fractions $\frac{1}{6}$ and $\frac{1}{3\frac{1}{4}}$ is $\frac{1}{6}$, and the difference between $\frac{1}{3\frac{1}{4}}$ and $\frac{1}{2}$ is $\frac{1}{5}$.

I have referred the commencement of expansion in water to the point of congelation. But it now seems generally supposed that water is contracted into the smallest volume about five or

six

six degrees above zero, and, in its descent beyond this stationary limit, again undergoes a slight dilatation. I am disposed, however, to question the accuracy of a principle so discordant and anomalous. In fact, the experiments on which it is grounded, though somewhat varied in their plan, never give the true expansions of water, but only the differences between those expansions and the corresponding expansions of glass. Having filled a thin glass ball, terminating in a fine tubular stem, with distilled water, and cooled the whole down to the point of congelation, I plunged it into a large bath, whose temperature was four or five degrees above zero. The water in the stem sunk at first considerably, owing evidently to the dilatation of the glass, and, by consequence, the enlarged capacity of the ball; but it then rose a sensible space, which must be ascribed to the expansion of the water itself. In like manner, when the procedure is reversed, and the ball, heated up a few degrees, is plunged into a bath at the point of congelation, the water rises in the stem as the ball contracts, and then, by its own contraction, partially subsides. The dilatation of glass

glass by heat is indeed so very small, that in most cases it may be safely disregarded. But the rate with which water contracts is perpetually diminishing as the heat declines, and therefore, at some particular point, this effect is exactly counteracted by the opposite contraction of the glass, and beyond it the latter must predominate. Nor is it difficult to determine, at least theoretically, the position of that *minimum*, or limit of apparent condensation. Water expands about the 24th part of its bulk between freezing and boiling; and glass, in the same interval, expands longitudinally the 1200dth part, and consequently its dilatation, in all the three dimensions, must amount to the 400dth part of its whole volume.* The expansion of water that corresponds to any temperature $x$ is therefore denoted by $\frac{1}{24}\left(\frac{x}{100}\right)^2$, and that of glass by $\frac{x}{40,000}$. Equating these two expressions, we obtain $\frac{x^2}{24} = \frac{x}{4}$, and therefore $x = 6°$. This remarkable coincidence seems to dispel every shadow of doubt, and we may embrace it as an established fact, that the successive dilata-

* See Note XXXV.

tions

tions of water, counting from zero, are as the natural progression of numbers.

Having distinguished its elements, we are now prepared to investigate strictly the process of cooling in a water bath. But although those elements are fewer, their mutual relation is yet of a more intricate and complex kind. I shall therefore divide the problem into two branches: first, when the water bath is at the point of congelation; and secondly, when it has any intermediate temperature between that and boiling.

1. The simpler case is where the bath of immersion is kept on the verge of congelation. After a diligent comparison of several detached observations, I am inclined to estimate the rate of cooling to be five times greater at the boiling, than at the freezing, point. If $h$, therefore, denotes the temperature of the body in degrees of the centigrade scale, the two concurring powers of refrigeration are represented by 1, and $4\left(\frac{h}{100}\right)^2$ or $\frac{h^2}{1250}$. Consequently, the relation of the time is expressed by this simple differential equation

— $dh$

$-dh = hdt\left(\frac{1250 + h^2}{1250}\right)$. The complete integral is $t = \text{Log.}\frac{H}{h} - \frac{1}{2}\,\text{Log.}\,\frac{1250 + H^2}{1250 + h^2}$,* where H and $h$ denote the two limiting temperatures corresponding to the interval of time, $t$. Or, if we introduce the proper coefficient, we shall have, in minutes, for the time of the cooling of a six inch ball, $\frac{1}{3}\,(\text{Log.}\frac{H}{h}\ \ \frac{1}{2}\,\text{Log.}\,\frac{1250 + H^2}{1250 + h^2})$, the first two figures after the logarithmic point being reckoned integers. Thus, towards the commencement of the scale, a metallic surface plunged in water cools 30 times faster, and a vitreous surface 14 times faster, than in common air. This refrigerating energy, however considerable it may appear, is yet much inferior to what we should expect from the comparative density and capacity of water, since it contains at least 500 times more heat than an equal volume of air.

2. When the water bath stands at any intermediate degree of temperature between its extreme limits, the problem becomes far more intricate. Let $h$ denote the heat of the bath, then the conjoined refrigerating energies will be repre-

* See Note XXXVI.

sented

sented by 1 and $\frac{h^2 - h'^2}{1250}$, whence the differential equation $dt = -\frac{dh}{h - h'}\left(\frac{1250}{1250 + h^2 - h'^2}\right)$. By resolving it into factors, we obtain this integral, $t = -\text{HLog.}\, h - h' + \frac{1}{2}\,\text{HLog.}\, 1250 - h'^2 + h^2 - \frac{h'}{(1250 - h'^2)^{\frac{1}{2}}}$ arc tang. $\frac{h}{(1250 - h'^2)^{\frac{1}{2}}}$ — Const.* Wherefore, putting $1250 - h'^2 = a^2$, the correct integral will be in common logarithms, $t = \frac{1}{3}\left(\text{Log.}\,\frac{H - h'}{h - h'} - \frac{1}{2}\,\text{Log.}\,\frac{a^2 + H^2}{a^2 + h^2} - M\,\frac{h'}{a}\left(\text{arc tang.}\,\frac{H}{a} - \text{arc tang.}\,\frac{h}{a}\right)\right)$; which expresses in minutes the time required for the cooling a six inch ball, M signifying the modulus, and the first two figures being reckoned integers. When $h'^2$ exceeds 1250, $a^2$ becomes negative, and consequently $a$, an impossible quantity. The last member of the integral, on this supposition, involves impossibles: in other words, the integration becomes impracticable by circular parts, but may be effected by the help of logarithms. Put $h'^2 - 1250 = \alpha^2$, and the complete integral will be $t = \frac{1}{3}\left(\text{Log.}\,\frac{H - h'}{h - h'} - \frac{1}{2}\,\text{Log.}\,\frac{H^2 - \alpha^2}{h^2 - \alpha^2} + \frac{h'}{2\alpha}\left(\text{Log.}\,\frac{H + \alpha}{H - \alpha} - \text{Log.}\,\frac{h + \alpha}{h - \alpha}\right)\right)$.

* See Note XXXVII.

It

It is obvious, that either of these formulæ will likewise apply when the ball immersed is colder than the bath. But I will not stop to remark all the varieties and modifications which they include. I must notice one case however, in which the expression becomes greatly simplified: it is when $h'^2 = 1250$, or $h' = 35^\circ\frac{1}{3}$, which corresponds almost exactly to blood-heat. On that supposition, $a$ or $\alpha$ vanishes, and the formula, dropping its last member, passes into $t = \frac{1}{2}$ (Log. $\frac{H - 35\frac{1}{3}}{h - 35\frac{1}{3}}$ — Log. $\frac{H}{h}$).—

Such are the combined principles which determine the refrigeration of a body surrounded by a fluid mass, whether air or water; and the same mode of investigation might be extended to other gases and liquids. Internal agitation gives prodigious activity to the circulation of the heat thus communicated. But exclusive of such accidental causes of acceleration, there is a constant and regular operation, by which its subsequent diffusion is chiefly carried on through the interior of the fluid. This results from the actual migration of the heated particles, which, being expanded, and

 therefore

therefore specifically lighter, endeavour continually to mount upwards, and assume their respective gradations. Nor is the buoyant force exerted only in a vertical direction; it has an evident tendency to generate lateral motions, since the heated portions of the fluid, by spreading out, are enabled to approach nearer the surface. Hence these quickly dispose themselves in horizontal strata, according to their respective degrees of temperature. This fact is distinctly perceived in a room with a stove, for the air is always warmer near the ceiling than above the floor. In a crowded theatre, the heat feels the most oppressive in the upper tire of boxes. Such familiar observations induced the ancients to ascribe to fire, whether apparent or in a latent form, a principle of *levity*; which, with some modifications, appears even at present not to be wholly rejected by philosophers.

But the distribution of heat by the efforts of ascension, is most conspicuous in liquids, which, from their comparatively ponderous qualities, are the least subject to external derangement or agitation. A vessel full of water is quickly heated

from

from below; but tedious and ineffectual is the attempt to direct the communication downwards. Yet considered abstractly from the species of matter with which it chances to be combined, heat is entirely passive, and neither betrays any natural reluctance to descend, or any inherent disposition to recede from the earth. A bar of iron will be heated almost equally soon, whether the upper or the lower end be thrust into the fire; and the slight difference of effect is occasioned by a stream of hot air, which always rises and glides along the sides of the bar. The horizontal diffusion of heat through fluids, and its graduated arrangement, are produced and moulded by the various progressive expansion of the affected particles. If cold water be gently poured on the surface of hot water, it will immediately sink to the bottom, without being sensibly warmed in its descent. Or if, by means of a long funnel, hot water be introduced at the bottom of a deep vessel which contains cold water, it will instantly spread and rise to the top. In a few seconds, the gradation of temperature at equal heights will in either case be the same. The tendency of those extreme temperatures to a mutual approximation, through the medium of the ordinary mode of communication, is comparatively very slow and imperfect: while the surface wastes its heat profusely, the bottom acquires only a small degree of increase. Nor, in a mass of water thus unequally heated, do the temperatures of the successive strata form an arithmetical series. I find, that the centre of the vessel always partakes more of the quality of the bottom than of that of the top, having only one third part of the whole excess of heat. The gradation of temperature, reckoning upwards, follows, therefore, an accelerating progression. This curious fact is explained by the increasing dilatations which water acquires from equal additions of heat; for the efforts of the heated portions of the fluid to ascend are thence continually invigorated in the higher temperatures. It is quite otherwise with alcohol, and, I presume, with oil: the centre of their containing vessel gives very nearly the mean temperature.

If the fluid has considerable depth, the difference of temperature between the successive strata must be proportionally small, and consequently the

the communication of heat through the mass will be diminished on a double account; the power of transmission being enfeebled, and the length of passage at the same time extended. A large collected body of water will, therefore, acquire or discharge heat, almost solely from the action of external impressions, which, according to their relative quality, will cause the particles immediately affected either to rise or to descend. Hence, the bottom of a very deep pool is always excessively cold; for the atmospheric influences are modified or diverted in their effects by the laws of statics. When the air becomes colder, the superficial particles of water, being chilled and of course condensed, sink downwards; but when it grows warmer, the particles which it comes to touch thence receiving heat and expansion, continue suspended at the surface. It is not, therefore, the mean temperature of the climate which is thus imparted: every change to warmth is spent on the upper stratum, while every transition to cold penetrates to the bottom; which suffers all the rigours of winter, without ever feeling the impression of the summer's heat. Nor is the peculiar effect counterbalanced, in any sensible degree, by the operation of other causes. Part only of the sun's rays strike the surface of the water, and this during a small portion of the year. But they are intercepted in their progress, and absorbed by the fluid mass; and though the bottom must continually receive heat from the bowels of the earth, yet the communication, being made through such slow conducting materials, the supply derived from that source is comparatively insignificant.

This remarkable phænomenon is strikingly exemplified in the lakes of Switzerland, whose vast depth is proportioned to the stupendous altitude of their encircling mountains. It appears from the careful observations of Saussure, that the bottoms of those majestic basins, whether situate in the lower plains, or embosomed in the region of the upper Alps, are almost equally cold, being only three or four degrees above the point of congelation. But the mean temperature of the ground over which the principal chain of lakes extends, is between 10 and 11 degrees. I found the heat exactly 10°, of a fine spring which

gushed

gushed up in a meadow below the romantic little town of Schweitz, and therefore not much above the level of the branching lake of the Forest Cantons. I likewise examined the temperature of water drawn by a pump from some depth in the charming isle of St. Pierre, seated in the lake of Bienne, about 200 feet above that of Geneva, and celebrated for affording a temporary retreat to the eloquent enthusiast Rousseau: it was 10½°.

The case is entirely altered in wide seas, that have an easy communication with the ocean. The tides and various currents which agitate the mass of waters, intermix the several strata, and produce an equal diffusion of heat. Thus, the same diligent and accurate naturalist discovered, at the depth of three hundred fathoms, the temperature of the gulf of Genoa, to be 15 degrees; which is precisely the average measure corresponding to that parallel of latitude.

The instrument which Saussure employed for making those observations, consisted of a strong coarse thermometer, surrounded by a considerable thickness of slow conducting materials. It was generally let down in the evening, and drawn up again next day. But notwithstanding its long continuance under water, this instrument could not give the full and correct result. However slow to receive impressions, it would evidently be affected during its ascent, especially by the length of the track which it described. We have observed already that, at low temperatures, the refrigerating action of water is nearly doubled, with the velocity of a foot in three seconds or of 20 feet in a minute. Motion through the fluid has, therefore, the same effect to cool a body as simple immersion during the time in which the passage would be performed with a velocity of 20 feet each minute, or 200 fathoms in an hour. But the depth of the lake of Geneva was 600 fathoms, at the place where the experiment was made; and consequently the thermometer would be as much affected in travelling through such a prodigious column, as if, in addition to the time of its ascent, it had remained suspended at the stratum of mean temperature for the space of three hours. This computed interval is so very considerable in comparison with the whole time of immersion, that it could not fail to have a material

terial influence in abridging the primary effect. I am hence strongly inclined to believe, that the bottoms of those profound lakes are always on the verge of freezing, or perhaps somewhat below it; for the superficial water is capable, even several degrees under that point, of resisting the process of congelation, and may consequently descend impregnated with their excess of cold. Nor is it impossible but the beds of such vast collections of fresh water are incrusted with banks of perpetual ice—a sort of subaqueous glaciers.*

Of the same nature is the curious fact which occurs in deep capacious reservoirs of stagnant air. But, from the nice mobility of that fluid, the phænomenon is here exhibited on a more contracted scale. It is not observed in narrow vales, enclosed by towering heights; because every breeze which sweeps over those summits will rouze the air from below, and invert or renew its internal arrangement. The permanent and intense coldness of the lower strata can only take place in profound caverns, open yet sheltered, anb

* See Note XXXVIII.

either

either perpendicular or gently inclined. Nor is there the same limit to the refrigerating intensity, as in the case of an aqueous accumulation. The mild air of summer hangs motionless at the mouth of the pit; but, in winter, the superior air, cooled many degrees perhaps under freezing, continually precipitates itself to the bottom.

I had occasion to witness the effects of this natural process, during an excursion through the mineral countries of the North, in the months of August and September 1799.—The famous Swedish mine of Dannemora, which furnishes the richest iron-ore in the world, presents an immense excavation, perhaps two or three hundred feet deep, and of still greater width. At the time time when I viewed it, the usual labours were suspended, for the construction of some masonry and other indispensable repairs. The bottom seemed full of water, in which were floating huge blocks of ice, or rather snow that had been soaked with humidity and again congealed. But the temperature of the ground, or that of water pumped up in the neighbourhood, I found amount to 7½ degrees.—The silver mine of Kongsberg,

Kongsberg, situate not far from Christiania, in Norway, has, for its main shaft, a frightful open cavern, perhaps three hundred feet deep, and thirty feet wide; in which the descent is made by ladders, resting against cross beams, without any platforms. The bottom is covered with perpetual snow, although the mean temperature corresponding to that latitude and elevation, is 6½.

A phænomenon of a similar kind, but far more curious and striking, occurs in the milder climate of France. It is a subterraneous glacier implanted at the bottom of a very deep cave in the centre of a forest adjoining the village of Beaune, which stands on the small river Doubs, about six leagues below Besançon, on the verge of the extensive chain of Mount Jura. The mouth is forty-five feet wide; and, after a long and steep descent, you enter a hall one hundred feet high; thence, by a slanting ladder of forty feet, you reach the chamber, which contains the congealed group. This consists of vast stalactites of solid ice, pendant from the roof, and nearly joining other branches, that shoot up from below.— Their origin is easily traced; for the snow which falls

falls in winter into the spacious hall, melts away during summer, and, percolating through the crevices of the rock, as it slowly trickles into the lower cavity, is arrested and consolidated by the action of the cold air.

Such are the singular effects of the principle by which the heated portions of a fluid invariably seek to occupy, or endeavour to maintain, an elevated situation. But this disposition to mount upwards has besides an indirect influence to promote the discharge of heat from the surface of a body immersed in a fluid, by causing a diffusive vertical motion, which limits and restrains the accumulation of a warm atmosphere: for it is evident that the refrigerating power of the medium, being always determined by its difference of temperature, would soon decline, if there were not some regular process by which the particles are removed from its vicinity as fast as they become affected, and a perpetual circulation thus kept up within the fluid mass. This buoyant tendency, however, performs yet a more important and extensive office. By its single operation, are fluids chiefly distinguished from solids

in

in their mode of transferring heat. But the subject requires some closer examination.—

It was shown that, besides the vibratory energy which is peculiar to the gases, and the regressive motion, which is common both to them and liquids; there is, in all fluid substances, a constant force exerted in discharging heat, similar to what obtains in solids. But the conducting power of a solid is not determined merely by the nature of the component materials; it is modified essentially by the space of transmission, being always inversely as the length of the communicating column. A body, plunged in a fluid, discharges its heat, however, with the same profusion, whatever be the extent of the surrounding mass: and if the medium of immersion be considerably contracted, this alteration, so far from accelerating the dispersive effect, will sensibly retard it. Heat is, therefore, conveyed from the body in the same manner as if this were enveloped with a solid crust of a certain determinate thickness; from whose exterior surface it is quickly absorbed, and thence transfused through the general fluid.

Suppose

Suppose the medium which encircles a hot ball to become suddenly fixed and solid. At first, the heat will flow copiously; but, as it advances and spreads, the current will afterwards gradually relax. The difference of temperature between the successive concentric shells into which the conducting mass may be distinguished, and consequently the measure of igneous transmission, must evidently diminish in proportion to the growing extent of penetration. When this influence has, therefore, acquired a wide extension, its subsequent communication will be extremely slow and languid. The consumption of heat in producing such diffusion, is comparatively small; but the rate of cooling must thenceforth be scarcely perceptible.—Let the concrete mass be now dissolved and restored to its original mobility. The hot portions of fluid will ascend on all sides, and quickly desert the vicinity of the ball, which will thus be left in the same condition as before. But it will not be entirely abandoned by its warm encircling atmosphere: for the particles of the contiguous film, in grazing along the surface of the ball, must experience such continued

nued obstruction as would very soon extinguish their acquired motion; and this effect would, from the mutual and intimate connexion that subsists, be communicated to the next particles, and by them to those which are immediately adjacent, till it seized the whole of the surrounding stratum or shell, to some limited depth. The shell thus affected, if not absolutely stagnant, will yet ascend only with the very slow progress to which corresponds a resistance equivalent to its buoyant force or quantity of heat. Beyond it the general mass of fluid, as fast as it receives the impression, will rise upwards with unimpaired mobility. As the intensity of the heat declines, the stratum contiguous to the ball will shift still more tardily; but the ascent of the rest of the fluid will likewise become proportionally slower. The reciprocal relations continue thus unaltered, and consequently the stagnating or obstructed atmosphere must, in every case, have the same thickness.

To elucidate this argument more fully, let A (fig. 27) be a point on the surface of the heated body, which is immersed in a fluid, AB the radial

radial extent of the medium, and AD the temperature at its origin; then, the mass being supposed in a concrete state, the slanting line DB will determine the temperature that corresponds to any given distance. But the rate of dispersion through the range of matter will be as the difference between the successive ordinates, and consequently as the tangent of the very small angle DBA. When the fluid is released, however, from consolidation, it will stream upwards, and transport its contained heat LBC, leaving only the superficial arrested shell AC. The rate of igneous communication will, therefore, be now represented by the oblique line DC; being increased in the proportion of AC to AB. But, beyond the stagnating atmosphere, the temperature of the fluid must still, in some degree, be affected; for the heat which is continually deposited at C, will not be instantaneously diffused and carried away. The limit of that atmosphere will not precisely reach to C, but to a point H (fig. 28), somewhat nearer, and FH will denote the temperature of the ulterior extending medium. This circulating portion of the fluid, though

though very slightly warmed, is yet enabled, by its considerable breadth and the celerity of its ascent, to transport the heat as fast as it is received. If the heat of the source be reduced to AE, the corresponding feeble temperature of the buoyant column will be expressed by GH. To preserve an equilibrium, therefore, the constant dissipation of heat must, in every case, be proportional to its original intensity. But as the warmth of the circulating mass declines, it will rise more slowly. This diminished effect, however, is exactly counterbalanced by the corresponding increased extent of the buoyant column; for the warmth must evidently penetrate farther into the fluid, if the celerity of the ascent be retarded. Thus, when the temperature of the body immersed is denoted by AD, the rising portion of the fluid, having the gentle warmth FH, will reach only to I; but, after that temperature has declined to E, the dispersive column, with the feeble heat GH, will exend to H. Though these exterior masses, therefore, mount upwards with different velocities, their energy of consumption will be respectively as FH and GH, or as AD and

AE, the central intensities of the heat. Hence the thickness AH of the stagnating atmosphere will always remain constant. What this thickness actually is, it would be difficult to ascertain with strict accuracy; but, from the comparison of a few facts which shall afterwards be recited, I infer that, in the case of air, the extreme boundary is not half an inch distant from the surface. In water, the separation is evidently much less.

I have employed straight lines to represent the gradations of temperature, merely for the sake of simplicity. But, as nature rejects all violent and sudden transitions, the exposition just given requires some restriction. The detained contiguous stratum is certainly not abruptly terminated; and the several films of which it consists are not each in the same degree stationary, those situate more remote from the surface having evidently a greater laxity and space for shifting. Nor is the origin of the buoyant column precisely limited: the passage from rest to motion, though not prolonged, is yet effected by regular shades. The portion of the fluid next the stagnating shell must

must still suffer some retardation. Instead of the compound lines DFM and EGN, the series of temperatures will be defined by similar curves, bending quickly towards the axis or assymptote AB, and then gliding beside it with a continued approach. But these considerations will not alter materially the preceding deductions; and I chose rather to avoid introducing at first a complication of views which would only embarras our reasonings.

## CHAPTER XVIII.

WE have investigated, and have explained at some length, the various circumstances which affect the progress of the cooling of bodies immersed in any species of fluid. We have considered their enclosing boundary, whether vitreous or metallic, as a mere physical surface; and have supposed the heat to be continually supplied from the internal mass, as fast at least as its consumption requires. Should the latter condition not obtain, it is obvious that the process will be proportionally retarded. Thus, if a thin hot ball, containing air only, be plunged in water; owing to the defective communication from the centre, it will not cool with the rapidity which might otherwise be expected. A similar effect takes place when the interior is filled with loose spongy substances, such as hair, wool, or feathers. But incidents of this sort occur so rarely, that they may be well overlooked. The case is more frequent and of much greater consequence, where the

the surface of a body is defended by a covering of slow conducting materials. The process of refrigeration is then retarded in proportion to the thickness of the exterior coat. On this principle depends the manifest utility of clothing, whether natural or artificial, in checking the too profuse dissipation of animal warmth.

A fluid of such extreme rarity as air, if confined round a heated body, must, like those spongy substances, have a decided influence to retard the operation of cooling. And this property is most distinctly perceived, though on a very limited scale. If a series of hollow cylindrical vessels be constructed of very thin brass, to fit into one another like a nest of boxes; the first or smallest filled with boiling water, and with a fine thermometer inserted, being enclosed in each of the rest consecutively, according to the order of their width, and kept equally separate from the sides and bottom, by resting against protuberant points or a slender checquered ring: on plunging the canister, with its adapted case, in a tub of water, the rate of cooling will be found, at every successive trial, regularly to diminish, till the space of

 inter-

intercluded air comes to exceed a quarter of an inch, when the effect will be reduced to about a sixteenth part. Beyond this limit, scarcely any farther decrease is observed, there now being room sufficient to allow that active fluid to develope its mobility, which fully compensates for the increasing distance of communication. A limit so narrow must evidently preclude the great majority of instances that would occur. The property of confined air to retard the progress of cooling is, therefore, founded on a principle not quite obvious, and not hitherto explained. By employing a series of concentric cases, or *septa*, this effect is wonderfully heightened. Yet a subject in itself so curious, and of such vast importance in the œconomy of heat, has been generally overlooked, or only treated in a vague and superficial manner. As I purpose to consider it with some attention, I shall, for the sake of clearness, divide it into three branches: 1. when the surface of the internal canister and its several cases are metallic: 2. when those surfaces are all painted, or consist of glass: and 3, when they are composed partially of both sorts.

1. When

1. When all the surfaces by which the included or exterior air is bounded, are metallic.—The pulsatory communication of heat, being here so much attenuated at each succeeding act of discharge and reception, may, in the general investigation of the problem, be fairly rejected. In fact, the first case would, by that process, acquire only the 72d part of the heat of the internal surface, and a second case would receive merely the 64th part of this small quantity, or the 4608th part of the whole. I shall assume, at least at the outset, that the rate of transmission is exactly uniform, and its intensity proportional to the difference of temperature; which must be very nearly true, when the central heat is moderate, and not affected by any sensible accumulative energies.

Suppose the canister so large that its surface may be regarded as equal to that of the exterior case, which is separated from it only by a narrow space. After an equilibrium is attained, the case will receive and discharge heat exactly in the same proportion; it must, therefore, be just as much hotter than the external, as it is colder

 than

than the included, air. But this confined portion will have evidently the mean temperature of its bounding sides. Consequently, reckoning the heat of the room as a standard, the temperature of the outer case must be equal to half the difference between itself and the temperature of the canister, or equal to one third of this whole quantity. Hence the canister, under the shelter of its case, will cool three times slower than if it were exposed naked.—Thus, when the central heat is 30°, that of the exterior surface will be 10°, and their arithmetical mean, or 20°, will be the temperature of the confined stratum of air. Therefore, the rate of internal communication, which cools the one surface in the same degree as it heats the other, will be as 10°, or equal to the discharge into the free surrounding atmosphere.

Imagine a second case to be now added. The mean temperature of the air which that contains is equal to its difference from the mean temperature of the air included within the first case; and either of these measures is equal to half the excess of the central heat above this last mean. Hence the outmost case will have only one fifth part of the

the temperature of the canister; and consequently, by the intervention of a double case, its rate of cooling is diminished five times.—For the sake of illustration, let the temperature of the central mass be 25°; then that of the first case will be 15°, and that of the second case 5°; the mean temperature of the inner stratum of air will be 20°, and that of the outer one 10°: the surface of the canister will regularly discharge a portion of heat as 5°; the next *septum* will receive and deliver the same to the contiguous air; and the external case will absorb this portion, and finally discharge it into the air of the apartment.

Pursuing the same mode of reasoning, it would be easy to show that, with three concentric cases, the canister would cool seven times more slowly; and with four such cases, nine times more slowly. In general, the degree of diminution is equal to double the number of cases increased by one, or the number of surfaces concerned: it is hence represented by the progression of the odd numbers, 3, 5, 7, 9, 11, 13, &c.

This result will appear sufficiently accurate, when the canister is of considerable size, and the

cases

cases not too widely disjoined; for instance, if the canister exceeds a foot in diameter, and the intervals between the cases are each of them not more than half an inch. But it is not difficult to obtain a rigorous solution, applicable even where the extent of the succeeding cases or *septa* is most rapidly progressive. With this view, I shall borrow the assistance of elementary geometry. Let the perpendicular lines AB and CD (fig. 29) denote the temperature of the canister, and that of its surrounding case; join BD, and produce it to meet AC in E; and make CF to AF as the surface of the canister is to that of the case, and draw FG: then will FG express the mean temperature of the included air, for DI : BH :: CF : AF, and the modifying effects of the opposite surfaces will obviously be proportional to their extent. But the exterior case will in the same degree receive and expend its heat; wherefore, DC = DI, and consequently CE = FC. The construction is thus manifest. Hence the rate of cooling is diminished in the ratio of AB to BH, or AE to AF. For example, suppose a cylindrical canister of three inches in diameter is enclosed within

within a similar case of four inches: then AF : FC :: 16 : 9, and AE = 16 + 9 + 9 = 34; consequently, the rate of cooling is reduced to $\frac{16}{34}$ or $\frac{8}{17}$, being rather less than the half. In general, if $a$ denote the diameter of the canister, $b$ that of its case, $\frac{b^2}{b^2 + 2a^2}$ will express the diminished rate of cooling. The value of this fraction, when $a$ is nearly equal to $b$, will, it is plain, approach to $\frac{1}{3}$, the same as what was obtained at first.

Conceive an exterior case to be applied. Let AB and CD (fig. 30), as before, represent the temperatures of the canister and the inner case; join BD, and produce it to meet AC in G. Make CH : AH, as the surface of the canister is to that of the inner case; and HI will represent the mean temperature of the intercluded stratum of air: let CK = CH, and KL will denote the mean temperature of the second stratum of air; make HE : CK, as the surface of the inner is to that of the outer case, and EF will express the temperature of this extreme surface; and EG being equal to EK, the rate of cooling will be diminished in the ratio of AG to AH. The reason of this procedure is apparent from the foregoing investigation,

tion, and the result is easily determined numerically. Thus, suppose the diameters of the canister and its two cases to be respectively 2, 3, and 4 inches; then, AH = 9, HC = CK = 4, and KE = EG = $\frac{9}{16} \times 4 = 2\frac{1}{4}$; whence, AG = $21\frac{1}{2}$, and the fraction $\frac{9}{21\frac{1}{2}}$, or $\frac{18}{43}$, marks how much slower the compound apparatus cools. In general, if those successive diameters be denoted by $a$, $b$, $c$, $d$, $e$, &c.; then will the diminished rate of cooling be expressed by $\frac{b^2}{b^2 + 2a^2 + \frac{b^2}{c^2} \cdot 2a^2 + \frac{b^2}{d^2} \cdot 2a^2 + \frac{b^2}{e^2} \cdot 2a^2}$ &c. or, perhaps more simply, thus,

$$\frac{b^2}{b^2 + 2a^2\left(1 + \frac{b^2}{c^2} + \frac{b^2}{d^2} + \frac{b^2}{e^2} \text{ \&c.}\right)}.$$

This formula plainly comprehends our first deductions. It would be superfluous to insist longer on the mode of analysis; but the theory is very satisfactorily confirmed by actual observations.

## EXPERIMENT LIII.

A cylindrical canister of planished tin, two inches in diameter and equal height, filled with boiling water, took 117′ to cool, from 30° to 10°; but enclosed within a similar canister of four inches

inches in diameter, it required 176′ to make the same descent. Another cylindrical canister of four inches, and which took 156′ to cool from 20° to 10°, required 356′ when cased with a similar one of five inches; yet, the interval being filled with flour, the effect was performed in 324′. And a square canister of three inches, that cools from 20° to 10° in 117′, took 335′ to perform the same effect, after it was enclosed within two similar cases of four and five inches.

These facts correspond with remarkable precision. Thus, in the first, $117 : 176 :: 16 : 16 + 4 + 4 = 1 : 1.5$; in the second, $156 : 356 :: 25 : 25 + 16 + 16 = 1 : 2.28$; and in the third, $117' : 335' :: 16 : 16 + 18\left(1 + \frac{16}{25}\right) = 1 : 2.845$.

In all the preceding investigations, it is assumed that the rate of cooling continues uniform. This postulate may indeed be admitted without material error in the lower temperatures; but when the central heat is intense, a very perceptible aberration will arise. It may, therefore, be desirable to obtain a solution exempt from any restriction.—When only a single case is interposed, the heat which flows from the canister must evidently divide itself into three successive

and

and equal portions: it is first discharged into the confined shell of air, thence received by the case, and lastly thrown from this into the general atmosphere. Put (fig. 29) AB $= h$, BH $= y$, and DI or DC $= x$; and let the surface of the canister be to that of its surrounding case as $m$ to $n$. Then the heat discharged from each point of the canister is denoted by $y\ (50 + y)$, and that received by each point of the case is, for a similar reason, denoted by $x\ (50 + x)$; therefore, $50\,my + my^2 = 50\,nx + nx^2$. But AB $=$ BH $+$ AH, or $h = y + 2x$; whence, by elimination, we obtain the quadratic, $x^2 - \frac{100\,m + 50\,n + 4\,mh}{4\,m - n}\,x = -\,m\cdot h.\,\frac{50 + h}{4m - n}$. And the value of $x$ being found, the diminished rate of cooling will be expressed by $\frac{nx}{mh}\left(\frac{50 + x}{50 + h}\right)$. When $4\,m$ exceeds $n$, the equation will have two roots, though the larger one is precluded by the nature of the problem.—An example will elucidate the application of the formulæ. Suppose a tin ball of three inches diameter, filled with water of the heat of 75°, is surrounded at a regular distance by a spherical cap of four inches: then, $m : n = 9 : 6$, $h = 75°$, and the equation becomes $x^2 - 220\,x$

$= - 4218.75$. Hence, $= 21.224$ or $198.776$, of which the first only can be admitted; wherefore the diminished rate of cooling, or $\frac{nx}{mh}$ $\left(\frac{50 + x}{50 + h}\right)$, is $\frac{24186.53}{84375} = \frac{43}{150}$, or $\frac{2}{7}$ nearly.

If $4m = n$, the quadratic becomes defective, and degenerates into a simple equation. On this hypothesis, $(100\,m + 50\,n + 4\,mh)\,x = mh\,(50 + h)$, and $x = \frac{mh\,(50 + h)}{m\,(100 + 4h) + 50\,n}$. Such is the value of the expression when the diameter of the case is double that of its included canister.----Suppose, for illustration, the former to be four inches, and the latter two, the heat of the mass being 50 degrees. The value of $x$ is consequently 10°, and comparative rate of cooling $= \frac{12}{25}$. But the result was confirmed by experiment: for, between 55° and 45°, the simple canister took 14′; while enclosed within the case, it required 29′ to make the same descent.

Resuming the general quadratic, it will be found that, $x\,(4m - n) = 50\,m + 25\,n + 2\,mh - \sqrt{(2500\,m\,(m + n) + 625\,n^2 + mn\,(150\,h + h^2))}$. If the canister be very large, the ratio of $m$ to $n$ may be considered as a ratio of equality,

this

this complex formula will become, by substitution and division, $3x = 75 + 2h - \sqrt{(5625 + 150h + h^2)}$. But the part affected by the radical is evidently quadrable, being equal to $75 + h$; consequently $x = \frac{1}{3}h$, which perfectly agrees with our first conclusion. Hence the diminished rate of cooling will be denoted by the fraction $\frac{1}{3}\left(\frac{50 + \frac{1}{3}h}{50 + h}\right)$, or $\frac{1}{3}\left(\frac{150 + h}{150 + 3h}\right)$. Therefore, in the higher temperatures, the exterior case has more efficacy to retard the process. Thus, while the extreme limit is $\frac{1}{3}$, the rate of cooling which corresponds to $50°$ is $\frac{2}{9}$, and that which corresponds to $100°$ is $\frac{5}{27}$.

The strict mode of solution which we have given for a single case, might, without much effort, be extended to any number of concentric *septa*; but the formulæ would become so complicated and fatiguing, that it seems better to rest satisfied with the former approximations. Or, if more accuracy be required for the higher temperatures, we may have recourse to the first construction. Thus, suppose the canister and its surrounding cases to have severally four, five, and six inches in diameter: then, AG = 25 + 32 +

$+\frac{25}{36} \times 32 = 79\frac{2}{9}$, and EG $= 11\frac{1}{9}$; consequently, if the heat of the canister be 50°, AG : EG :: 713 : 100 :: 50 : $7\frac{1}{80}$, or FE, the temperature of the outmost case. Hence the rate of cooling is assigned by the compound ratio of $100 \times 713 \times 16 : 57\frac{1}{80} \times 100 \times 36$, or 1140800 : 213245, or is about $5\frac{1}{3}$ times slower. The extreme limit is $\frac{79\frac{2}{9}}{25}$, or $3\frac{1}{6}$ times slower. It would be easy to exemplify the mode of calculation in other instances of a more involved nature.

2. The next division of the problem is, where the canister and its surrounding cases are painted or vitreous. This condition will be found to alter materially the proportion of the result. When two such surfaces, with unequal degrees of heat, are made to front each other, they will not, like metallic plates, act the same as if they were quite insulated; but must, by their pulsatory energies, exert a mutual influence to accelerate the progress towards an equilibrium. If their visual magnitude be very considerable, or their extent great in comparison of their distance, almost the whole of those opposite dispersive pulsations will be intercepted and received on both sides. But

 with

with a moderate difference of temperature, the vibratory discharge constitutes very nearly the half of the ordinary measure of communication. Therefore the vitreous or painted surfaces must emit or absorb heat one half faster than if they were removed beyond each other's sphere of action, but accompanied by the same intercluded atmosphere.

If the one surface be completely encompassed by the other, it is evident, that the exterior will receive all the diverging pulsations; and if the interior be not disproportionately small, it must, in its turn, intercept those which are reciprocally convergent. Let then AB, and CD (fig. 31) denote the temperature of a canister, and that of its surrounding case: make CF to AF as the surface of the former is to that of the latter; and GF will, as before, express the mean temperature of the interjacent air. If the internal surfaces were unconnected, each point of the former would discharge a portion of heat as BH, and each point of the latter would receive a portion of heat as DI. But, in consequence of their communication, the vibratory impressions are mutually doubled, and therefore

therefore the cumulative effect is augmented by one half. The case receives from the canister a hot pulse equal to $\frac{1}{2}$ DI, and the canister intercepts from the case a cold pulse equal to $\frac{1}{2}$ BH. Hence CD $= \frac{1}{2}$ DI; for the flow of heat outwards from the case into free space must keep pace with its increased absorption. Consequently CE $= \frac{1}{2}$ CF, and the construction is obvious. The rate of cooling will now be represented by $\frac{CE}{AE} \times \frac{AF}{FC}$, or $\frac{3\,AF}{2\,AE}$; the dispersion from the outer case being compounded of the intensity and relative quantity of surface.—The same conclusion may be derived somewhat differently; for the temperature of the canister is reduced in the ratio of AE to AF, or from AB to BH, which, by the pulsatory reaction of the adjacent case, has its energy increased one half.----Suppose, by way of illustration, the canister and its case to have four and five inches in diameter: then, AE $= 25 + 16 + 24$, and the diminished rate of cooling $= \frac{37\frac{1}{2}}{65} = \frac{15}{26}$, which agrees with observation.

Let a second case be now added. The previous

 construction will remain the same as when those *septa* consisted of planished tin. AB, CD, and EF (fig. 32), will represent the temperatures of the canister and its successive cases; HC will be to AH as the surface of the canister is to that of the inner case, and KE will be to CK as the surface of that inner case is to the surface of the outer case; HC will be equal to CK, since the internal *septum* is confined alike on both sides; but EG will be equal to $\frac{3}{2}$ KE, for the terminating surface discharges its heat into free space with only two thirds of the whole internal energy. The diminished rate of cooling is hence denoted by $\frac{3\,AH}{2\,AG}$.

In general, if $a, b, c, d, \ldots q$, denote the diameters of the canister and its series of cases, the diminished rate of cooling will be expressed by the formula, $\frac{3b^2}{2b^2 + a^2\left(4+4\frac{b^2}{c^2} + 4\frac{b^2}{d^2} \cdots 5\frac{b^2}{q^2}\right)}$. When the canister is extremely large, the quantities $a^2$, $b^2$, $c^2$, &c. may be considered as having to each other a ratio of equality, and if $n$ denote the number of cases, the formula will be reduced

to

to $\frac{3}{3+4n}$; thus constituting the progression $\frac{3}{7}$, $\frac{3}{11}$, $\frac{3}{15}$, &c.

Where the canister is enclosed within a single case, its diminished rate of cooling, we have seen, is expressed by $\frac{3}{2} \times \frac{AF}{AE}$. But if the case be continually enlarged, FC, and consequently FE, will proportionally decrease. Therefore the value of the fraction $\frac{AF}{FE}$ will thus always approach to unity, its ultimate limit. Hence, on the supposition that the case is of vast or unbounded extent, the central canister, instead of having its refrigeration in some degree impeded, would actually cool one half faster than if it were suspended in absolute free space. This paradoxical conclusion is utterly inadmissible, and implies a latent inaccuracy, which has become apparent on being magnified. In fact, we assumed that the canister intercepted *all* the cold reflex vibrations sent convergent from the case. But this position is not strictly correct, except when there is but little interval between the canister and its case, and the latter has a circular form well adapted to concentrate its reacting impressions. If the canister be much reduced, it will intercept only a small part of the wide pulsations. But as these are twice more powerful in front than their mean intensity, the intercepted impression may be estimated proportional to double the surface of the canister. The rate of the canister's cooling, instead of being expressed by $\frac{3}{2}$ AF, is therefore represented by AF + FC. But this acceleration is evidently reciprocal; or the power of the case to disperse heat externally is to that of receiving it from internal communication, as AF to AC. Hence, to maintain the equilibrium of absorption and discharge, FC must be to CE, or the internal to the external difference of temperature, in the same ratio. Consequently AC (fig. 33), being a mean proportional between AF and AE, the modified rate of cooling is denoted by $\frac{AE}{AC}$, or $\frac{AC}{AF}$. The refrigeration of the canister is thus always something retarded by the influence of the surrounding case, though it approaches fast to its ultimate limit of equality.

3. The last branch into which the problem divides itself is that, where vitreous and metallic surfaces are promiscuously combined. But as such

such possible combinations must evidently be very numerous, I shall select only their principal varieties.

Suppose a painted canister is included within a bright tin case. If the reflective power of the internal surface of the case were absolutely complete, the progress of refrigeration would be exactly the same as if the canister had a metallic lustre; for the discharge of heat by pulsation would then be rendered altogether abortive, being constantly sent back from the case to its source, and there re-absorbed. The effect would thus be comparatively much greater than in any of the preceding instances. However, the defective reflection, or partial absorption, of the tin, sensibly modifies the result. It is obvious, that the mean temperature of the intercluded air will be determined in the same manner as before. But, while a polished metallic surface emits nine parts of heat, a painted one disperses sixteen. Of the sixteen parts, therefore, which the canister is capable of discharging, no more than ten prove really effective, the additional part only being absorbed by the inner surface of the case. Hence the temperature of the exterior surface must be somewhat greater than the mean internal difference, to enable it to disperse its invigorated accessions of heat into free space. Therefore, ID : DC (fig. 29), or FC : CE, as 9 : 10; and if A denote the surface of the canister, B that of its case, the reduced temperature of the former will be = $\frac{B}{B + \frac{19}{9} A}$, or $\frac{9 B}{9 B + 19 A}$. Therefore the diminished rate of cooling is denoted by $\frac{10}{16} \left(\frac{9 B}{9 B + 19 A}\right)$, or $\frac{90 B}{144 B + 304 A}$. Hence, A being considered as equal to B, a very large painted canister will cool almost five times slower, when surrounded by a case of planished tin.

Let the position be now reversed, the surface of the canister being clear, and its exterior case painted on both sides. This case will, therefore, absorb at its inner surface ten parts of heat, of which the canister makes an effective discharge; but, with the same difference of temperature, it would disperse sixteen parts into the free external atmosphere. Hence DC = $\frac{10}{16}$ ID, or CE = $\frac{10}{16}$ FC; consequently the diminished rate of cooling is denoted by $\frac{10}{9} \left(\frac{B}{B + \frac{26}{16} A}\right)$, or $\frac{80 B}{72 B + 117 A}$.

Suppose the case to be painted on one side only. If the canister is bright, and the surface fronting it is painted, then will FC = $\frac{10}{16}$ CE, and the diminished rate of cooling will be denoted by $\frac{10}{9}\left(\frac{B}{B+\frac{26}{16}A}\right) = \frac{50\,B}{45\,B + 117\,A}$. But, if conversely, the canister is painted, and the inside of the case metallic, we shall have CE = $\frac{10}{16}$ FC, and the diminished rate of cooling expressed by $\frac{10}{16}\left(\frac{B}{B+\frac{26}{16}A}\right) = \frac{10\,B}{16\,B + 26\,A}$.

These formulæ are not strictly applicable, except the elevation of temperature be small, and the metallic surfaces have an elaborate polish. In practice it will be more accurate to substitute the numbers 8, 5 and 6 respectively, instead of 16, 9 and 10. The several formulæ will then become: 1, $\frac{15\,B}{20\,B + 44\,A}$; 2, $\frac{24\,B}{20\,B + 35\,A}$; 3, $\frac{18\,B}{15\,B + 35\,A}$; and 4, $\frac{3\,B}{4\,B + 7\,A}$. The first denotes the diminished rate of cooling with a painted canister and a polished case; the second, that corresponding to a bright canister within a painted case; the third and fourth express the retardation produced by alternating painted surfaces.

It

It would be superfluous to prosecute this subject any farther. The examples which have been chosen are sufficient to explain the varying mode of investigation. When several cases are employed, alternately vitreous and metallic, the effect is nearly the same as if they were all metallic, but the general influence will depend chiefly on the quality of the outmost surface. For the same reason, the vitreous or painted surfaces will have much less power to retard the process of cooling, when they lie adjacent, than when they are interspersed.

In all the preceding deductions, the exterior atmosphere is supposed to be in a state of perfect calm. But if the surrounding air is agitated, this motion will evidently modify the result; for the surface exposed, then discharging its heat more profusely than before, will, with a smaller difference of temperature, be enabled to eject the continual accessions from the interior. The ratio of EG to EK (fig. 30), must thus be diminished in proportion as the celerity of the stream increases. Suppose, for instance, that the action of the wind were sufficient to accelerate $n$ times the refrigeration

refrigeration of the surface on which it is immediately exerted; then the reduced rate at which the compound apparatus must cool, if the several septa are metallic, is denoted by

$$\frac{B}{B + A\left(2\,\frac{B}{B} + 2\,\frac{B}{C} + 2\,\frac{B}{D}, \&c. \ldots \frac{1+n}{n}\,\frac{B}{Q}\right)}.$$

The painting of the outmost surface would occasion a very trifling difference of effect, the co-efficient of the last term being only changed from $\frac{1+n}{n}$ to $\frac{2+n}{n}$.

Let the canister and its cases be considered as of equal extent; since $n = 8$, when the wind is extremely vehement, the retardation of cooling that corresponds to 1, 2, 3, &c. cases, is respectively $2\frac{1}{8}$, $4\frac{1}{8}$, $6\frac{1}{8}$, &c. instead of 3, 5, 7, &c. and if the outmost surface be painted, this retardation will be denoted by $2\frac{1}{9}$, $4\frac{1}{9}$, $6\frac{1}{9}$, &c. Suppose the canister to be included in a single case only, and both of them painted: the rate of cooling will be $= \frac{3}{2}\left(\frac{B}{B + \frac{2n+3}{2n}A}\right) = \frac{3B}{2B + \frac{2n+3}{n}A}$.

If we consider B as equal to A, this fraction will very nearly approach to unity; or the canister, screened

screened by its case, will not cool faster in the strongest wind, than if it had stood uncovered in still air.

With equal facility may be determined the progress of refrigeration which obtains on the immersion of the apparatus in a liquid mass. Since the discharge of heat by external pulsation is now precluded, the nature of the extreme boundary will have no influence whatever on the measure of effect. This result must depend almost entirely on the quality, the position, and the number of the interior surfaces.—Let the canister, with its series of metallic cases, be plunged in a bath of water: then, $GE = \frac{1}{30} EK$ (fig. 30), for heat was found to be dispersed from a surface of tin about thirty times faster in water than in air; consequently, the diminished rate of cooling will be expressed by

$$\frac{B}{B + 2A\left(\frac{B}{B} + \frac{B}{C} + \frac{B}{D}, \&c. + \frac{31}{60}\cdot\frac{B}{Q}\right)},$$ or more simply, $$\frac{B}{B + A\left(2 + 2\frac{B}{C} + 2\frac{B}{D}, \&c. + \frac{B}{Q}\right)}.$$ If A, B, C, D, &c. be all esteemed equal, the corresponding rates of cooling will be successively denoted by

by the fractions $\frac{1}{2}$, $\frac{1}{4}$, $\frac{1}{6}$, &c. instead of $\frac{1}{3}$, $\frac{1}{5}$, $\frac{1}{7}$, &c. which express the degrees of effect that would take place in atmospheric air. Suppose the canister and its cases are now vitreous or painted: then, $GE = \frac{1}{30} \times \frac{2}{3} KE = \frac{1}{45} KE$, and the diminished rate of cooling =

$$\frac{3\,B}{2\,B + 2\,A\left(2\,\frac{B}{B} + 2\,\frac{B}{C}, \text{ \&c.} + \frac{46}{45}\,\frac{B}{Q}\right)},$$ or very nearly

$$\frac{3\,B}{2\,B + 2\,A\left(2\,\frac{B}{B} + 2\,\frac{B}{C} + 2\,\frac{B}{D}, \text{ \&c.} + \frac{B}{Q}\right)}.$$ Let A, B, C, D, &c. as before, be reckoned equal, and the successive rates of cooling will be denoted by the fractions $\frac{3}{4}$, $\frac{3}{8}$, $\frac{3}{12}$, &c. instead of, $\frac{3}{7}$, $\frac{3}{11}$, $\frac{3}{15}$, &c. which represent the corresponding effects in air.

Resuming the strict formulæ, and supposing the canister is surrounded only by a single case, the rate of cooling will, for metallic surfaces, be $\frac{30\,B}{30\,B + 31\,A}$, and, for painted or vitreous surfaces, $\frac{135\,B}{90\,B + 92\,A}$. But the different results are best illustrasted by contrast. A cylindrical canister of planished tin, three inches in diameter and height, and which in still air takes 117′ to cool from

from 20° to 10°, would require 249′, if included regularly within a similar cylinder of four inches; but only 185′, if the whole were immersed in a tub of water. The same canister, when painted, would, in a close room, cool in 6 ′, or, surrounded with its case, likewise painted, in 98′; and both plunged in water, would take only 64′.

In all these examples, the canister and its several cases are regularly separated from each other by intervals of half an inch. If the divisions approach nearer, their effect soon becomes altered; for the successive strata of intercluded air, as they diminish in thickness, lose in some degree their internal mobility, and begin passively to transmit heat like a solid mass. When the terminating surfaces mutually approximate, not only is the fluidity of the thin shells of air proportionally cramped, but the power of communication is likewise invigorated by the shortness of the passage and consequently the quicker gradation of temperature. On both these accounts, therefore, the quantity of transmission will increase with most rapid progress as the *septa* contract their

their limits. Thus a cylindrical tin canister of three inches in diameter and height, placed within a similar one of four inches, will cool about one sixtieth part faster, if shifted from its position in the middle to a quarter of an inch from the bottom; and nearly one ninetieth part still faster, when advanced only an eighth of an inch from that boundary. Hence we may compute, that a stratum of air one quarter of an inch thick, transmits through its substance about a sixth part of the heat which it is fitted to communicate in the ordinary mode, and if reduced to half this thickness, it will deliver nearly equal shares in both ways.

But to discover more accurately the progress of this internal transmission, I procured another intermediate cylinder of tin, with a moveable lid, and three inches and three quarters in diameter. The three inch canister inclosed within this, had its rate of cooling reduced to 7-15ths. But calculation gives $\frac{45}{97}$, the difference being only $\frac{4}{1455}$, and consequently the aberration or accelerating influence corresponding to an interval of three eighths of an inch, must be extremely small. The

The diameters of the cylinders are as 4 and 5, and their surfaces as 16 and 25. Therefore, the temperature of the internal canister being denoted by unit, that of the outer case is $\frac{7}{15} \times \frac{16}{25} = \frac{112}{375}$. The mutual difference is $\frac{263}{375}$, and hence the canister exceeds the temperature of the interjacent air by $\frac{25}{25+16} \times \frac{263}{375} = \frac{160\frac{15}{41}}{375}$. This fraction will express the ordinary measure of communication; but the actual discharge of heat is 7-15ths, or $\frac{175}{375}$, and therefore $\frac{14\frac{26}{41}}{375}$, or about the twelfth part of the whole is conveyed away through the stratum of air by passive transmission.

When the intermediate cylinder was included within the four inch one, their interval being only the eighth part of an inch, the deviation appeared now to have most rapidly increased. The rate of cooling, instead of 17-47ths, was only reduced to 17-26ths. The opposite surfaces being as 225 to 256, or very nearly as 15 to 17, the temperature of the exterior one is $\frac{17}{26} \times \frac{15}{17} = \frac{15}{20}$, and consequently the internal canister must exceed

exceed the temperature of the thin stratum of intercluded air by $\frac{11}{26} \times \frac{17}{32} = \frac{187}{832}$. This must denote the ordinary discharge of heat; but the real consumption is $\frac{17}{26} = \frac{544}{832}$, which is nearly triple the former. Therefore when the shell of air is only the eighth of an inch in thickness,---- of 31 parts of heat 10 are carried off by the general process and 21 by quiescent communication.

But the close proximity, and still more the partial contact, of the canister with its exterior case, has not merely a negative influence to diminish the retardation of cooling. It must actually accelerate the dispersion of heat, since, in effect, it occasions an artificial enlargement of surface. A tin canister of two inches square will cool one half slower, when planted in the centre of a similar one of four inches. But if it be made to touch three sides of the case, it will cool about three times faster than at first; for these sides, having the same temperature as the canister, and presenting twice its extent of surface, must double the refrigerating action, exclusive of the co-operation of the remaining sides, which will add at least one half more.

To produce their proper effect, therefore, it is requisite that the cases should be perfectly detached or insulated. The retardation of the process of cooling depends entirely on the coldness of the external surface. But metals conduct heat so freely, that even a partial contact might be sufficient to cause an almost equal diffusion. If a round tin vessel, of a broad and rather flat shape, have a cap fitted at each end capable of being drawn out to different small distances, the rate of cooling will continue very nearly the same through all the gradations, from the position of absolute contiguity till the circular plates are separated by an interval of perhaps three quarters of an inch. The narrow rims, embracing the canister, rapidly abstract heat, and convey it to the prolonged boundaries. We hence see the defect of the ordinary form of pots with double lids, designed for culinary purposes.*

* See Note XXXIX.

CHAP-

## CHAPTER XIX.

THE progress of research now leads me to describe the construction of the *Photometer*; an instrument of uncommon delicacy, which I have invented to measure the force or density of light. Light and heat were proved to be only different states of the same identical substance. When the lucid particles are intercepted and absorbed, the corresponding accession of warmth uniformly discovers itself by a proportional dilatation. If the calorific action of light could, therefore, be separately educed,—if a receptive material were selected capable of large and regular expansions,—if the heat thus collected could be next defended from the influence of variable extraneous impressions,—and if its subsequent dispersion were impeded by some constant obstructing cause; the intensity of the luminous stream would then be always expressed by the permanent elevation of temperature which it must occasion. Such are the principles on which

which I have composed the photometer. By persevering trials, I have successively improved its mode of execution; and this instrument has at length acquired that simplicity, if not elegance, of form, which seems to mark the limit of perfection.

The mensuration of the degrees of light has been much later cultivated than other parts of the science of optics. The celebrated Huygens threw out some hypothetical ideas relative to this sort of inquiry: but the first that appears to have considered it with attention, was Marie, a Capuchin friar, who, at the close of the seventeenth century, wrote a small book expressly on the subject. This obscure person, however, from his slender acquaintance with mathematics, committed a radical mistake. It was reserved for the sagacity of the ingenious Bouguer to unfold the principles of that interesting branch of optical science. His elementary work came out in 1729, and exhibited the rare union of profound geometrical skill with habits of nice observation. In the sequel, it was greatly enlarged, and republished by Lacaille from the author's manuscript, two

two years after his death, in 1760. At this very time, the famous Lambert printed his *Photomctria*; and from the Greek, a language so smooth and flexible, composed that term, which has since been adopted into the several dialects of Europe. The treatise of the Swiss philosopher embraces a wider compass, and displays the various resources of his ardent and fertile genius. We have only to regret that the facts on which his conclusions depend are but too often lame and inaccurate. The deficiency of experiment he frequently attempted to supply by geometrical analogies, and thus carried to a blameable excess the method of interpolation, which, if managed with caution and address, is of such vast utility in the prosecution of all physical inquiries. Whether from poverty or a love of singularity, that extraordinary man contantly declined the assistance of able artists, and sought to draw every thing from his own resources and individual efforts. This circumstance lessens in a most essential degree the value of his deductions.

Since that period, photometry has attracted little notice, and has made very small advances.

Some alterations, perhaps improvements, have indeed been introduced into the fundamental apparatus employed by Bouguer: but I cannot help remarking that machinery of such a complex nature is by no means entitled to the name of photometer. Each observation performed by it is really a distinct process of experiment, and which requires dexterity and skill in the operator.—The principle of the construction is abundantly simple. Though the eye is not fitted to judge the proportional force of different lights, it can distinguish, in many cases with great precision, when two similar surfaces, presented together, are equally illuminated. But as the lucid particles are darted in right lines, they must spread uniformly, and hence their density will diminish in the duplicate ratio of their distance. From the respective situations, therefore, of the centres of divergency, when the contrasted surfaces become equally bright, we may easily compute their relative degrees of intensity. Yet a most material objection still remains: the apparatus admits no certain standard of comparison. Even the force of the sun itself, at the same altitude

tude and in the same climate, is subject to considerable variation. And how could we fix the power of a lamp or a candle? What a variety of undefinable circumstances intermingle their influence! The size and brilliancy of the flame are not determined merely by the diameter of the wick, nor by the consumption of the oil or tallow.

It has always appeared to me that a preferable mode of estimating the force of light, if not too much attenuated, might be derived from the measure of its calorific effect. But I considered it as more eligible to determine the initial rate of action, than to mark the extreme limit of progressive accumulation, which seemed liable to be deranged by irregular influences. For that purpose, I procured a most delicate mercurial thermometer, with a long stem bearing only a few degrees, but with a bore of such fine and equal calibre, that, the bulb being half an inch in diameter, each degree had an extent of nearly two inches. The bulb was covered with a coat of China ink, and detached about an inch and a half from the scale, close to the extremity of which the stem was encircled by a broad ring of cork: to this was

 occasionally

occasionally adapted a short tube of glass, somewhat tapering, and more than an inch wide, having both ends ground square, and to the outer one a circular piece of thin clear glass cemented. The thermometer turned firmly about a pivot that passed through the scale, and was fixed to a stand. The instrument, thus prepared, being placed directly in a sun-beam suddenly admitted through a hole above an inch wide made in the window-shutter, the number of seconds which elapsed during the ascent of the mercury through the space of one degree was carefully observed by a stop-watch. The expansions appeared at first remarkably regular and uniform, but afterwards continually diminished, as the heat now collected began copiously to disperse. It is manifest that the power of illumination, being proportioned to the measure or celerity of action, must be in the reciprocal ratio of the time required for producing a given effect. I thus deduced the comparative intensities of the solar rays at various angles of obliquity, and consequently affected by various lengths of passage through the atmosphere; and I experienced the singular satisfaction

of

of finding them to correspond almost exactly with the results which Bouguer had obtained in a very different way. Those observations were performed in the year 1791, and I designed to prosecute the inquiry more extensively, if my attention had not been soon diverted to other objects.

Having conceived the idea of a differential thermometer, composed of two counteracting balls, and which I had successfully employed in the construction of an hygrometer, it seemed an easy and obvious step to accommodate the same principle to the mensuration of light. Yet I will freely confess, that this reflection did not occur to me till more than two years after, and was then suggested merely by a lucky incident. We are indeed every day surprized that the simplest thoughts should escape us. Perhaps the spirit of discovery consists not in the power to create new combinations, but in the quickness to perceive each anomalous appearance, and the patience to trace out its remotest consequences. I well knew the great force of the solar beams, yet, in common with most people, I had formed an erroneous and very defective estimate of the intensity of the light

light which is reflected from the sky. In the one case, the impression is concentrated on the retina; in the other, it is enfeebled by diffusion: and we are thence grossly deceived respecting the combined action of these indirect and scattered rays.

The correct performance of the differential thermometer indispensably requires that, in the state of rest, it should, under all the vicissitudes of temperature to which it may be exposed, constantly point at the zero of its scale. The air contained in either ball ought, therefore, in a similar degree to have its elasticity augmented or diminished by the corresponding accession or abstraction of the same measure of heat. But this effect might be disturbed by the unequal action of the interposed fluid, in modifying the constitution of the opposite portions of imprisoned gas. Alcohol, which answers so well for thermometers, is totally unfit for the present object; since air attracts it eagerly, and acquires, from its union, a great expansive force. If this dissolving process were indeed alike exerted at both extremities of the column, the equilibrium would still be maintained.

maintained. But the cylindrical reservoir evidently presents a much broader surface than the slender bore, and consequently the absorption or deposition of humidity, produced by the change of temperature, will be far more rapid and complete in contact with the former. Mercury, therefore, appears at first to be the fluid best adapted. On closer inspection, however, we find it utterly inapplicable to the purpose; for not only its great ponderosity would occasion small and crowded divisions, but its motion is irregular and subsultory, as it suffers much obstruction in passing through narrow tubes, and forces its way by successive wide starts.* The fixed vegetable oils might seem to possess the property wanted: yet even these are not altogether exempt from atmospheric absorption; and hence, after some time, they grow clammy, and adhere to the sides of the tube. Deliquiate potash promises to combine every requisite; for it moves with facility, and soon balances, by its reaction, the dissolvent power of the confined air. But a certain degree of concentration is necessary towards the permanence of this equilibrium, at which any moderate

* See Note XL.

rate addition of heat, while it increases the aërial attraction, produces likewise an equal and contrary effect, by augmenting the adhesion of the humidity to its alkaline basis. After I had ascertained that limit, and found the differential thermometer, when so constructed, to stand for several days invariably at the same point, I was surprized to remark a considerable alteration on suspending it out of doors in the shade. I could hardly impute this anomaly to the sudden change of temperature, which must besides have been very small. The instrument, being designed for an hygrometer, had the one ball clear, and the other covered with a bit of cambric. It therefore occurred to me, that the derangement might be occasioned by the unequal influence of light. I immediately wrapped a piece of black silk above the cambric, and was delighted to observe the aberration much increased. Having transferred the silk to the clear ball, the disturbing effect was not only extinguished, but a still greater one produced in the opposite direction. This incident happened in the summer of 1797; and forgetting for a while the hygrometer, I now directed

directed all my efforts to the application of the primary instrument to the measuring of light. I left one of the balls naked, and coated the other with China ink: the ordinary impression in open air was very considerable, and I soon learnt to proportion the several parts to the general extent of action. This simple construction is not, however, sufficient to form an exact photometer; and though fitted to mark each prominent gradation of light, it will not always exhibit regular and proportionate effects. The accession of heat, during any given time, is evidently as the number of lucid particles that are absorbed by the black ball; but the degree of its accumulation must depend on the slowness with which it is again dispersed. If the heat were uniformly conducted away at a rate proportioned to the excess of temperature, nothing could be wanted to the perfection of the instrument. But this continual discharge is produced and determined by the agitation, whether excited or pre-existent, in the atmosphere. In the case of the hygrometer, the condition of the surrounding fluid with respect to motion or rest, has no influence whatever in

modifying

modifying or disturbing the proper effect. The augmented energy of action may there hasten the term of equilibrium, yet it cannot have the smallest tendency to depress that ultimate position; for if the quicker renewal of air on the humid surface, by animating the process of evaporation, accelerates the abstraction of heat, the greater frequency of contact likewise multiplies, in the same proportion, the successive counterbalancing deposits. But the photometer, in its naked form, and where no such balance obtains, must be materially affected by the degree of stillness or commotion of the atmosphere, and would therefore, under the same force of incident light, give most variable and uncertain results. Perhaps the most remarkable circumstance is, that the extremes should not be more widely separated. In the strongest wind which I have observed, the impression of a clear meridian sun upon the dark ball still amounted to near the fifth part of what the same action produced in calm weather.

But those perplexing irregularities are entirely removed, and the desired equipoise most effectually

tually introduced, by inclosing the instrument within a glass case. This was the first step which I made, and it answered my sanguine expectations. Though still imperfectly acquainted with the theory of its operation, I soon convinced myself that the width of the case was of very small consequence. I therefore proceeded with confidence in the execution of the plan; and it was highly gratifying at first to witness the nice coincidence of different instruments constructed after different proportions.

My next object was to render the photometer conveniently portable, by reducing its size, and giving it such a form of construction as might exempt it from every possible chance of accidental derangement. Nor should I forget to mention the elegant improvement of employing black enamel, instead of merely painting the ball with China ink.—But it was still a desideratum, to procure a liquid capable of retaining its colour and of preserving its equilibrium with the contiguous portions of air. Deliquiate potash, which takes a fine tinge from archil, and once seemed to answer so well the required conditions, was found

found quite unable to resist severer trials. Exposed to a strong light, it gradually deposited its colouring matter, while at the same time it continually shifted its place nearer the ball into which the small reservoir opened; thus obviously betraying all the effects which proceed from the absorption of oxygen. I then thought of filling the balls with hydrogene gas, instead of common air; imagining, by this expedient, to exclude entirely the presence of oxygen. The experiment promised complete success, and I flattered myself that I had at last discovered the object of my anxious research. But I was again deceived: after some time, I perceived with pain that the purple tinge slowly faded away; in the space of a few months, the alkaline liquor turned almost limpid, and retreated considerably from its first position.* I now began to despair of the possibility of finding a coloured fluid capable of withstanding the combined action of air and light. Rejecting the medium of hydrogene gas, I contented myself with employing sulphuric acid in its pure limpid state. A permanent stability was

* See Note XLI.

thus

thus obtained; but, while that liquid formed no contrast with the tube in which it moved, the instrument appeared to labour under a sensible defect. I tried the effect of tinging the sulphuric acid with carmine, and at last, beyond all expectation, I succeeded to the utmost of my wishes. The colour is a beautiful crimson, perfectly durable; at first indeed it suffers a very slight impression, but it never afterwards in the smallest degree is affected by the operation of light and confined atmospheric air. I have several instruments which have undergone no alteration whatever, since the time that they were made, in the spring of 1800. Hence I may safely consider the photometer as arrived at its ultimate state of improvement. I shall, therefore, proceed to describe the simplest and most eligible mode of constructing this curious instrument.—

The first business is to prepare the coloured liquor; which is done, by adding a pinch of carmine to sulphuric acid in a small phial with a ground stopper. The crimson dye is soon communicated, though it continues for several days to grow more intense. The process of solution is hastened by occasionally stirring the phial, but the application of heat must be avoided, as it makes the acid to char the colouring powder and assume an inky turbid appearance. The undissolved flocules will subside to the bottom, or the liquid may be decanted off into another phial.—The glass tubes are next selected, as regular as possible, from 4 to 8 inches long, and about 3-20ths of an inch thick; or as slender as those employed for thermometers, but with a much wider bore. This, in one tube, must have from the 40th to the 60th part of an inch in diameter, and an exact calibre, at least not differing by a fiftieth part between both its extremities. To the end of it, a small piece of black enamel is attached, and blown into an opaque ball, from four to six-teenths of an inch in diameter. The corresponding tube may have its bore of the same, or rather a greater, width, but its uniformity is not at all essential. Near the extremity it is swelled out into a thin cylinder, almost one-tenth of an inch wide, and from three to six-tenths long; the inner cavity only being enlarged, without altering the

the exterior regularity of the tube. The short bit of glass where this cylinder terminates, is now blown into a thin pellucid ball, as nearly of the size of the former as the eye can judge. The exact equality of the balls would be unattainable, and fortunately the theory of the instrument does not require it. When a dark and a bright object are viewed together, the latter, from an optical deception, appears always larger than the reality; and, for this reason, I prefer making the clear ball a slight degree smaller than the black one. The tubes are now cut to nearly equal lengths, and the end of each swelled out a little, to facilitate their junction. Close to the black ball, the tube is bent by the flame of a candle into a shoulder, such, that the root of the ball shall come into a line with the inner edge of the tube. This ball, being then warmed, the end of the tube is dipt into the acid liquor, and as much of it allowed to rise and flow into the cavity, as may be guessed sufficient to fill both tubes, excepting the cylinder. The two tubes are then, by the help of a blowpipe, solidly joined together in one straight piece, without leaving any knot or protuberance. About

 half

half an inch from the joining and nearer the cylinder, it is gently bent round by the flame of a candle, till the clear ball is brought to touch the tube three quarters of an inch directly below the black one. The instrument is next held in an oblique position, that the coloured liquor may collect at the bottom of the black ball, into which a few minute portions of air must, from time to time, be forced over, by heating the opposite ball with the hand. In this way, the interposed liquid will gradually be made to descend into the tube, and assume its proper place: and it should remain for a week or two in an inclined position, to let every particle drain out of the black ball. If any trail of fluid collects in rings within the bore, they are easily dispelled with a little dexterity and manipulation, which, though it would be difficult to describe, is most readily learnt and practised. The small cavity at the joining, facilitates the rectification, by affording the means of sending a globule of air in either direction. In fixing the zero of the scale, I set the instrument in a remote corner of the room, or partly close the window-shutters. When completely

pletely adjusted, the top of the coloured liquor, if held upright, should stand nearly opposite to the middle of the cylindrical reservoir.

In this state of preparation, the instrument is ready for being graduated. I therefore cover the clear ball and the contiguous part of the parallel tube with two or three folds of thin bibulous paper, which I moisten with pure water, to make it act as an hygrometer; and attach to the same tube a temporary scale, by means of a soft cement composed of bees-wax and rosin. A flat round piece of wood being provided with four or five pillars that screw into it, I fix to one of them the instrument erect, and dispose on either side two fine corresponding thermometers inverted and at the same height, the one having its bulb covered with wet bibulous paper. Then, taking about half a yard of flannel, I dry it as much as possible, without singeing, before a good fire, and rolling it up like a sleeve, I lap it loosely round the lower part of the pillars, and enclose the whole under a large bell-glass. The flannel powerfully absorbs moisture from the confined air, and creates an artificial

 dryness

dryness of 80 or 100 degrees. In the space of a quarter or half an hour, the full effect is produced, and the quantities being noted at two or three separate times, the mean results are adopted. The descent, measured by the temporary scale, being then augmented in the proportion of ten to the difference of the two thermometers, will give the length that corresponds to 100 photometric degrees.—After the standard instrument is constructed, others are thence graduated with the utmost ease; the first being planted in the centre, and the rest, with their temporary scales, stuck to the encircling pillars. For greater accuracy, the observation should be made in a room without a fire, or a screen ought to be interposed between the fire and the apparatus.

The slips of ivory intended for the scales are divided into equal parts, according to the respective size of the degrees, of which they should each contain from 100 to 150. The edges are filed down and chamfered, to fit easily between the parallel tubes; and they are secured in their place by a strong solution of isinglass, which

dries

dries into a thin glossy film, that adheres most firmly, and indeed answers better than any kind of cement.

The glass case is an essential part of the photometer. It should be thin and as translucid as possible, the top neatly rounded over and hermetically sealed, without leaving any knob or smoky tarnish. Its form ought to be a cylindrical, from six to eight-tenths of an inch wide; or such that, when fitted to the instrument and the one side touching the graduated branch, the opposite side should be separated from the balls by a space exceeding the twentieth, if not the tenth, part of an inch. The proper situation of the black ball is about three quarters of an inch below the top of the cylinder.

The instrument is fixed into a piece of ebony, by means of black sealing-wax. This must be done with great caution: a mortice, rather wide, is made close to the edge of the wood, and full three quarters of an inch deep; and the bending of the compound tube, where the scale leaves it bare, is gently heated and inserted, the hard cement being melted and poured round it. After this has become nearly cold, the surface is smoothed with the point of a hot iron. The glass case slides upon a shoulder of about half an inch long, to which it is fastened with soft cement. Below it, are three or four threads of a screw; then a narrow bead, and another similar screw. The outer case of ebony being screwed on, a small cap is likewise screwed at the end, for the sake both of protection and symmetry of appearance. When it is wanted to measure the intensity of light, the cap and exterior case being removed, the instrument is screwed to a circular bottom of two or three inches in diameter. Or, without using the stand, the case may be reversed and screwed to the lower end, and then serve to hold the whole in a perpendicular position. (See Plate VI. where fig. 35 exhibits the instrument as prepared for action, and fig. 36 represents it shut up in its case and fitted for the pocket.)

Such is the construction of the portable photometer; but, for more accurate and philosophical observations, I prefer a form somewhat different. The cylindrical reservoir is near a quarter of an inch longer, and both the balls are alike reflected,

and

and front each other at the same height. The glass case consists of a cylinder of about an inch and half wide, to which is applied and fastened by help of isinglass the larger segment of a clear globe, perhaps three inches in diameter, and such as commonly used for cutting into watch-glasses. The instrument, being cemented into the middle of a small piece of ebony or lignum vitæ, is, before the fixing of the spherical cap, let down and secured with soft cement to the inner extremity of the cylinder. The protruding screw fits into a broad circular bottom, which for greater safety may be loaded with lead. (See Plate VII.)

This mode of construction is not adapted for carriage, but it possesses other eminent advantages. The spherical shell can be selected of admirable evenness and transparency; and as the counteracting balls have the same elevation, they are entirely exempt from any irregularities that might arise, in experiments where great or sudden heat is concerned, from the disposition of the warm portions of air to arrange themselves in horizontal strata.

The theory of this instrument I have in a great measure

measure anticipated. It is evidently in no respect liable to be affected by any variations of temperature in the surrounding atmosphere. These, acting alike upon both balls, must always produce a mutual compensation. If the one ball, for example, grows warmer, the other will acquire heat exactly in the same degree; and therefore the intervening liquor, under equal and opposite elastic pressures, will still remain stationary.—But the accession of light destroys this equilibrium. The incident rays freely traverse the clear ball, without exciting any effect: they are, however, detained and absorbed at the surface of the black ball, and there, assuming a latent form, they act as heat. Hence the temperature of the black ball will continue to rise, till the increasing dispersion of heat, caused by the process of refrigeration, becomes equal to the regular supply derived from the incessant influx of light. When this expenditure of heat is then equivalent to its accession, an equipoise will again be produced. But it was proved, that the discharge of heat from a body is more copious very nearly as the excess of its temperature above that of the air in which it is immersed. And since this

this difference is supposed to be inconsiderable, and the surface affected is vitreous, the proportion must be almost rigorously exact. Therefore the heat finally acquired by the black ball, or the corresponding depression of the liquor in the tube, will measure the intensity of the incident light. This descent is at first pretty rapid; but it soon relaxes, and after the space of two or three minutes, it entirely ceases.

Such is the manner in which the instrument acts, when exposed naked to the influence of a calm atmosphere. Whatever has a tendency to retard the cooling of the sentient ball, without likewise affecting the other, must augment proportionally the measure of impression. If we could possibly substitute a black metallic surface, we should thereby double the quantity of effect; for the absorption of light continuing the same, its subsequent dispersion in the form of heat, being under similar circumstances twice as slow, must require a double excess of temperature to maintain the balance of receipt and expenditure. Hence, on covering the dark ball carefully with a bit of tin-

tin-foil, the photometric impression is nearly the same as before. In fact, though the bright surface reflects about one half of the incident light, it compensates for this profuse waste, by the almost total want of pulsatory dispersion, which constitutes in general the moiety of the consumption of heat. As the coat of tin grows tarnished by exposure, the action of light upon it increases, till the metal approaches the condition of an oxyd, and its power of refrigeration, now invigorated, counterbalances its augmented absorption. Gold, next to copper, is the metal that appears to have the feeblest reflection, and which is also the least subject to discolour or to suffer a partial oxydation. It might answer, therefore, to have the upper ball blown of clear glass, and afterwards gilt or enamelled with gold. But the advantages of this mode of construction do not correspond to its beauty. The impression made on that precious material is much less than we should expect; and though esteemed a perfect metal, it is yet subject to a slow oxydating process, which gradually deepens its tint, and consequently alters its power of absorption.

It

It was shown that the pulsatory discharge of heat is independent of all extraneous influence, and that the agitation of the atmosphere promotes the cooling of a surface, merely by increasing the other sources of expenditure. Hence, on a gilt ball, the sweeping wind has more power to abridge the impression of light, than on a ball of black glass. A moderate breeze which reduces the effect made on the former to the third part, will contract that produced on the latter to near a fifth.

The addition of a glass case is therefore rendered indispensable. It not only precludes all irregular action, but maintains, around the sentient part of the instrument, an atmosphere of perpetual calm. With the same force of incident light, the black ball must still rise to the same height above the temperature of its encircling medium. The case will evidently have some effect however to confine the heat received, and consequently to warm up the internal air. Therefore, equivalent to this excess, the temperature of the black ball must acquire an additional elevation: but the clear ball, being immersed in the same fluid,

fluid, must experience a similar effect, which will exactly counterbalance the former. The difference of temperature between the opposite balls thus continues unaltered; and neither has the size or shape of the case, nor the variable state of the exterior atmosphere with respect to rest or agitation, any influence whatever to derange or modify the results which are exhibited by the photometer. This distinguished property merits yet a fuller examination.—

Conceive the exterior case to be at once transparent and metallic. As it must finally discharge all the heat which is deposited on the surface of the black ball, it will first become proportionally warmer, and above this point of temperature the included air will acquire a corresponding elevation. But the clear ball must assume the temperature of its encircling medium, and consequently will obtain an equal degree of ascent. Hence the difference of calorific effect produced on the opposite balls, and which alone the instrument is calculated to mark, will remain still the same. And such would be the process, if the sentient ball were

were also metallic. Its vitreous quality, however, renders the operation of the case more complicated. The portion of heat emitted from the black ball by pulsation, amounting to one half of the whole, is not immediately or completely absorbed by the bounding surface, but thrown from side to side, till, after repeated reflections, it is partly received by the case, and partly restored to its ball or communicated to the other. Yet this secondary accession of heat, exerting the same effect on either ball, will raise both of them equally above their previous state of equilibrium. Therefore the absolute difference of temperature is not, in the slightest manner, altered.—Thus, suppose that the incident light is able to produce an effect of 111 degrees on the naked instrument, and that the affected internal zone of metal extending nearly opposite to the connected balls, is equal to thrice their surface. Of the vibratory discharge the eighth part, or the sixteenth of the whole measure of communication, will be absorbed by the metallic surface. The remaining 7-16ths will be repeatedly bandied from side to side, till it is spent on that surface and the included balls.

balls. At each reflection, the case will detain one-eighth, and each of the balls, one-sixth; or the residuum will be distributed in the proportion of six to the case, and eight to each of those balls. Hence the portion of heat restored to the black ball is $=\frac{7}{16} \times \frac{8}{22} = \frac{7}{44}$, and that deposited on the internal surface of the case is $= \frac{1}{16} + \frac{7}{16} \times \frac{6}{22} = \frac{2}{11}$. Therefore the refrigerating energy of the black ball is reduced to $\frac{37}{44}$, and consequently the incident light must excite an effect on it $= 111^\circ \times \frac{44}{37} = 132^\circ$. But the temperature of the included air exceeds that of the case by the sixth part of this quantity, or 22°; the mutual differences of temperature being always inversely as the opposite surfaces. And since the case receives internally $\frac{2}{11}$ of the vitreous emission, or $\frac{4}{11}$ of what would be discharged from a metallic surface, it must exceed the temperature of the external atmosphere by 30°, or by 22° increased in the proportion of 15 to 11. Thus the total impression made on the black ball amounts to 132 + 22 + 30, or 184°. But the clear ball receives, from the repeated reflection, an accession of heat equal to $\frac{7}{44}$, which corresponds to 21°. The combined

bined effect produced on this ball is therefore $= 21 + 22 + 30$, or $73^\circ$; but the difference of temperature which the instrument indicates is $184-73$, or $111^\circ$, being the same as before.

A diaphanous case of metal is evidently hypothetical, and I have introduced it merely for the sake of contrast and illustration. But a case of glass has likewise the property of occasioning an equal deviation in each of the connected balls. In raising the temperature of the included air, it does not change their relative state; and the opposite influence of its internal surface, to promote refrigeration by a reacting pulsatory energy, is exerted equally on them both. Let AF be to FC (fig. 33) as the effective part of the case is to the surface of the black ball: Make AF : AC :: AC : AE; draw the perpendiculars AB, FG, and CD, and join BE. If AB represents the temperature of the black ball, FG will denote that of the included air, and CD that of the case itself. Exclusive therefore of any reflex vibratory action, the effect marked by the instrument is BH, or BA—HA; being the very same as that which would obtain, if the case were entirely removed, but the

 atmosphere

atmosphere of immersion still heated up to the interior temperature FG. The vitreous surface however, sending a force of pulsation equal to DI, depresses the clear ball to its own temperature AO. But the same impression accelerates the cooling of the black ball, in the proportion of BH to BO. The intensity of light capable of supporting this increased profusion, instead of BH, is consequently now BO; which is also the difference between the temperatures, BA and OA, of the two balls. Thus the instrument continues to indicate the true calorific power of the incident rays.

Hence the clear ball is actually somewhat colder than the internal medium, having always the temperature of the enclosing vitreous surface. The refrigerating energy, however, of the black ball is not as BH, the excess of its temperature above that of the included air, but as BO. It is therefore the same as if the cooling were performed without auxiliary excitement, and the temperature of the fluid of immersion were reduced from AH to AO. By its vibratory reaction, the case in effect depresses this medium to its own standard.

Suppose

Suppose the force of light to be 90°, and the part of the case affected to be five times the surface of either ball. Then BH = 75°, HO = 15, and OA = 18°. The whole heat of the black ball is 108°, and that of the clear ball, or the case itself, 18°, leaving 90° to express, as before, the luminous energy.

If the case be contracted, it will become proportionally warmer, to enable it, from a smaller surface, to discharge the internal heat. The standard temperature will, therefore, rise in the same degree; but, both the balls being alike affected, their difference of elevation must remain unaltered. On the contrary, if the case be enlarged, or its dispersive energy be augmented by immersing it in water or exposing it to a current of air, it will grow colder, and the interior level will consequently subside, by an equal measure of descent. Nor will the thickness of the case have any tendency to alter the relation of the balls. It will only, by retarding the dispersion of the heat, raise the temperature of the confined air. And thus, from the happy combination of balancing principles, no extraneous influence, however

 powerful,

powerful, can at all affect the accuracy of the results.

But if the receptive ball is gilt, the instrument will be liable to a slight variation, according to the different size of the case. Since metal discharges heat only half as fast as glass, make AF to FC (fig. 34) as half the surface of either ball is to the effective portion of the case. And because the mutual action of the case and the gilt ball is eight times less than before, make FC to CE as AF is to AF + $\frac{1}{8}$ FC. But though the reflex pulsation of the case be eight times diminished, it will exert on the metallic surface an effect equal to the fourth part. Therefore, let HN = $\frac{1}{4}$ HO; and BN will denote the real intensity of the light, while BO will represent that which the instrument exhibits. This error, however, is in general very small, and, where the case has an extensive surface, it may be fairly neglected. Suppose this to be $n$ times the surface of the gilt ball; it would be easy to show that $\frac{BO}{BN} = \frac{8n + 4}{8n + 3}$, and consequently that the derangement is expressed by the fraction $\frac{1}{8n + 3}$.

I have

I have tacitly presumed that the case admits the whole of the incident rays. This indeed is not strictly true; yet if the glass be of a good quality, and drawn or blown to a proper degree of thinness, the loss which the light sustains by absorption during its passage, will be very inconsiderable. I seldom found such deficiency to exceed the twentieth part; and still less must be the portion of light detained by the clear ball.—There is also a small waste of power arising from the partial reflection or incomplete absorption of the rays by the glossy surface of the black ball. This could be prevented almost entirely by applying a coat of China ink; but as it likewise amounts only to about the twentieth part of the incident beam, and has in every instance the same proportion, it may be disregarded. These combined sources of dissipation will not cause a diminution of effect equal to the tenth part of the whole. Nor would this waste, were it ever so great, produce the slightest derangement in the relative operation of the instrument.*

* See Note XLII.

## CHAPTER XX.

I HAVE now detailed the construction, and explained at some length the theory, of the photometer. It yet remains for me to point out its application. I am disposed to regard it not only as an instrument of very considerable curiosity, but as calculated to give us correcter notions on a variety of interesting subjects, and as admirably fitted, by its extreme sensibility, to assist us in the prosecution of several philosophical enquiries. To the superior aid of instruments and artificial combinations does the science of physics owe its rapid advances. Before the invention of the thermometer in the beginning of the seventeenth century, men were accustomed to judge the different degrees of heat merely by their feelings. But the estimates thus formed were often highly exaggerated, and always vague and fallacious. The acquisition of that valuable instrument first introduced certainty and precision; and by detecting minute alterations of temperature

and

and changes of mutual relation, it has led to most important discoveries.—To measure the variable intensities of light promises ultimately similar advantages. It is that quickening energy which pervades the universe, and gladdens the face of nature: it is the emblem, and the perpetual fountain, of almost every joy and comfort that sweetens this feverish state of existence and renders life desirable. This Proteus of the material world is unceasingly varying its force, and changing its fugacious forms. We contemplate an extended scale of light in the vicissitude of day and night, in the revolutions of the seasons, and the unequal distribution of climate over the surface of the globe. The photometer exhibits distinctly the progress of illumination from the morning's dawn to the full vigour of noon, and thence its gradual decline, till evening has spread her sober mantle; it marks the growth of light from the winter solstice to height of summer, and its subsequent decay through the dusky shades of autumn; and it enables us to compare, with numerical accuracy, the brightness of distant countries,—the brilliant sky of Italy, for instance, with the murky air of Holland.

It is not my design at present, nor have I yet sufficient materials, to enter into the detail of such researches. I wish only to give some general ideas of the photometrical observations; reserving for another occasion the full statement and discussion of the results.—In the latitude of 56°, the direct impression of the sun at noon, during the summer solstice, amounts to 90 degrees; but it regularly declines, as his rays become more oblique. At the altitude of 17°, it is already reduced to the half; and at 3° above the horizon, the whole effect exceeds not one degree. In the same parallel of latitude, the greatest force of the solar beams, in the depth of winter, measures only 25 degrees. Their diminished vigour is evidently caused by the dispersion and absorption which they must suffer in their protracted slanting passage through the atmosphere. The law of decrease is likewise nearly that which has, from other principles, been assigned by the ingenious Bouguer; but, in this country at least, it is subject to some variation and uncertainty, from the imperfect clearness of our insular sky. Between a fourth and a fifth part of the light of the sun is lost in a vertical descent to the surface súrface of the earth : and as, even in the finest weather, a thin haze generally floats near the horizon, the successive waste corresponding to each equal number of aërial particles which a very oblique ray encounters in its track, will often amount to the third.

Of the quantity of indirect light which is reflected from the sky, we are apt to form a false estimate, in consequence of its being so much attenuated by diffusion. But though extremely fluctuating, it is often very considerable. In this climate, it may amount to 30 or 40 degrees in summer, and to 10 or 15 in winter. This secondary light is the most powerful when the sky is overspread with thin fleecy clouds; it is feeblest in two very different conditions, either when the rays are obstructed by a mass of thick congregated vapours, or the atmosphere is quite clear and of a pure azure tint. In mists and low fogs, the diminution of the light is comparatively small, it being then affected more by indistinctness than any want of intensity. This reflex illumination has its diurnal progress more extended and much slower than the force of the sunbeams. It commences

mences with early dawn, and in the fine season of the year, before the sun has yet emerged from the horizon, it measures five degrees. In a short space of time however, it is equalled and next surpassed by the rapid ascent of his resplendent orb.

In the higher regions of the atmosphere, the rays of the sun, not being impaired by such a length of passage, are more vigorous than at the surface of the earth; but the diffuse indirect light of the sky, as it is reflected from a rarer mass of air, is therefore proportionally feebler. It would be most interesting, to make observations of that kind on the lofty summits of the Alps or the Andes. The traveller who visits those elevated tracts, is struck with the dark hue of the azure expanse, through which, with some effort, he can, even during the day time, discern certain of the brighter stars. I once attempted, with very imperfect means, to compare the force of the sun on the top of the height situate on the west side of the pass of the Great St. Bernard, with that which is experienced in the plain below, at Martigny in the Valais. The nearer impression of the rays seemed about one-fourth

fourth part greater. I would recommend that spot as a commodious station for performing such experiments. It is besides a position famous in the annals both of ancient and modern war; and commands a noble prospect of Mont Blanc, rising awfully with his snowy flanks amidst a vast amphitheatre of mountains.

When the sky is obscured by a dense body of clouds, the darkness seems to be much increased in proportion to the obliquity of the solar rays. In summer, the photometer, placed in the open air at noon, seldom or never marks less than 10 degrees; but in some of those sable-shrouded days which, in this remote region, deform the winter,* I have repeatedly observed, that the whole effect, under similar circumstances, did hardly exceed even one degree. It might be curious to compare the extreme changes of light and darkness that frequently occur within the tropics; where a vast mass of pitchy clouds, rising charged with thunder, will suddenly overcast and obstruct the azure vault of heaven.

* ————" When the dragon womb,
" Of Stygian darkness spits her thickest gloom,
" And makes one blot of all the air."——

When

When the photometer stands insulated out of doors, it must evidently receive the rays that come from all sides. But set in the window, it can only feel the impression caused by a part of the sky. Unless it chance to front the quarter in which the sun shines, it will seldom indicate more than 15 degrees. On drawing the instrument back into the room, the effect will rapidly decrease; for the intensity of action is obviously proportional to the visual space subtended by the window. It requires near two degrees of light, to enable one to read or write with pleasure: a greater portion offends by its excessive glare, and a much smaller quantity fatigues and strains the eyes.

Placed in open air, the photometer is not only affected by the light sent from the sky, but also in some measure by what is reflected from the ground. This derangement is for the most part very small, and may easily be excluded altogether, by fixing a screen or circular horizontal rim about the glass case near the top of the scale. The reflection from a green field perhaps exceeds not the twentieth of the whole incidence; it however increases considerably as the colour inclines to white.

white. In a steady winter day, the photometer, standing on a piece of newly-ploughed ground composed of black mould, indicated 24 degrees: moved to a plot of russet grass, it marked 30; transported to a smooth sandy beach, it showed 33; and placing the instrument on a broad surface of snow, the effect rose to 44 degrees. Thus, snow almost doubles the force of impression, by joining a copious reflection. Its numerous facets, presented in every possible position, disperse the incident rays in all directions. Yet it does not reflect the whole, and about one-sixth is absorbed and lost. This observation completely disproves the notion sometimes entertained, that snow is possessed of a sort of phosphorescent quality, which enables it to emit light from its own substance.

The photometer affords a ready method of ascertaining the various degrees of transparency. Many substances not commonly reckoned diaphanous, are yet very pervious to light. White paper transmits the rays, irregularly indeed, but in such profusion, that it will serve as an occasional substitute for glass. Of 100 parts of the whole

incident

incident light, I find that cambric admits 80, and when wetted, 93. Vellum paper suffers 49 parts to pass through it, thin post 62; but, soaked in olive oil, the former will allow the passage of 80 parts, and the latter, that of 86. In penetrating through repeated folds, the quantity of transmission decreases after a geometrical progression. Thus, thin post rolled four times about the cylindrical case of the photometer, will reduce the density of the light received to 15, and vellum paper applied as often will diminish it to 6.— In traversing paper and other similar substances, the rays are not, therefore, sent directly through the supposed vacuities or pores. The wide dispersion which they suffer in the passage, proves clearly that they make their escape by some intricate tracks, and experience various deflections from the repulsive and alternate energies of the proximate matter. The addition of water or oil to the cambric or paper, forms a real chemical union, and bestows on the compound an intermediate character, more inclined however to the nature of a fluid. The moistened substance not only transmits the rays of light more copiously, but with

with a smaller deviation; or, it allows a larger proportion of them now to pursue their course, without being turned aside.

The intensity of the light reflected from any surface, is ascertained with almost equal ease. It is only requisite to provide a screen for excluding the sun's direct rays, or to place the instrument in a situation where it can only receive the reflected beam. By varying the angle of incidence, we may determine the series of corresponding effects. A number of important results will be thus obtained.

The photometer is usefully employed to measure the relative density of various artificial lights. It enables us to compare most accurately the power of the solar rays with the force of illumination produced by a candle, a taper, or a lamp. For this purpose, I use the second form of construction, where the two balls are oppositely reclined at the same height. A wax candle of 9-10ths of an inch in diameter was placed directly before the instrument, and its flame two inches distant from each ball. In the space of a few minutes, the photometer received an impression

sion of six degrees; and on drawing back the candle by successive steps, the effect, as we should expect, regularly decreased as the square of its distance. Thus, with an interval of three inches from the nearer surfaces of the balls, the action of the flame was reduced to $1\frac{1}{2}$ degrees; and consequently the sixth part of a degree would express its whole amount at the distance of a foot.—A tallow candle of the same diameter, but with a flame three times larger, produced, under similar circumstances, an impression only one-half more. Therefore the wax candle burnt with a flame twice as bright, or emitted light twice as much concentrated. The consumption of the tallow candle was at the rate of an inch every hour, that of the wax candle was an inch and quarter in the same time. Hence tallow yields proportionally more light than wax, but a light not so clear and condensed.

On placing the photometer immediately before a coal fire and 30 inches from it, the effect communicated by its dull reddish light amounted to eight degrees. But the luminous surface presented by the fire, deducting the dark parts of the

coal

coal and the bars of the grate, was equal to a circle of $6\frac{1}{2}$ inches in diameter. Consequently at the distance of a foot, the same impression would have been made by a circle of fire, whose diameter is $6\frac{1}{2} \times \frac{12}{30}$ or $2\frac{1}{5}$ inches. But the flame of the wax candle was equal to a circle of only 3-7ths of an inch in diameter, or very nearly six times smaller. Therefore if the flame of the candle had the same extent of surface as the fire, its effect would have been $\frac{1}{6} \times 36$, or six degrees. Thus the calorific energy of the light emitted by a coal fire is greater by a third part, than that of the brilliant condensed flame of a wax candle.—If the fire be very powerful, we might endanger the photometer by placing it too near. In that case, a large pane of glass should be interposed a few inches before the instrument, and about one-fifteenth part allowed for the waste by absorption.

It will hence be easy to compare the power of different lamps, and to determine the relative advantages of each particular mode of construction. This useful inquiry, however, I have not yet had a convenient opportunity of prosecuting. Yet I am rather inclined to believe, that, in point of

 œconomy,

œconomy, the difference is not very material. The light shed by the several inflammable substances with which we are acquainted, may vary in its purity and degree of concentration, but its absolute quantity appears always nearly proportioned to their rate of consumption.

Among the celestial bodies, a very wide diversity however obtains, with respect to their powers of illumination. The action of the flame of a wax candle, at the distance of a foot, did not exceed the sixth part of a degree; but the full impression of the solar rays, if not enfeebled by their passage through the atmosphère, would amount to 125 degrees. The surface of the flame was equal to a circle of 3-7ths of an inch in diameter, or the 28th part of its distance. The mean apparent diameter of the sun is four times less, for the subtense of the angle of 30′ 42″ corresponds to the 112th part of the radius. Consequently if the wax candle were removed to the distance of four feet, its flame would present the same visual magnitude as the sun himself; but the effect which it could then produce would be diminished 16 times, or would amount only to the 96th part of a degree.

Therefore

Therefore the light emitted by the sun is 96 × 125 or 12,000 times more intense than that of a wax candle; or if a portion of solar matter only 3-7ths of an inch in diameter could be transported to this planet, it would afford as much light as twelve thousand such candles.

The light of the moon has the opposite character of excessive debility. The action of her rays on the photometer is quite imperceptible; nor could I render it visible, even by collecting them in the focus of a large burning glass. But I was enabled to form some estimate, by an indirect mode of comparison. I selected a small table of logarithms on which I could barely read the figures, by the light of the full moon: on retiring gradually backwards from a wax candle set to burn in a darkened room, I found the figures now become indistinct, beyond the distance of 15 feet. The force of the light received from the candle must have been only the 1350th part of a degree, for $\frac{1}{6} \times \left(\frac{1}{15}\right)^2 = \frac{1}{6 \times 225} = \frac{1}{1350}$; and consequently if the flame had been contracted to the same apparent magnitude as the moon, this measure would have been diminished still 16 times

 more,

more, and hence reduced to the 21,600th part of a degree. But the illuminating power of the sun, at the same altitude, is 70 degrees, and therefore exceeds that of the moon, in the ratio of 70 × 21,600 to 1; or, in round numbers, it is *one hundred and fifty thousand times* greater.

This estimate is double what has been assigned by the celebrated Bouguer; and my respect for the conclusions of that able observer has induced me, where the limit was dubious, to lean more to the side of defect than of excess. If I have erred therefore, I presume it is in representing the lunar illumination rather too small than too large. But neither of these computations will agree with the current opinion, that the moon derives her light merely from the sun. In fact, if the moon reflected and dispersed in every direction the whole of the light which she receives from the sun, it would, before it reached us, be spread over the concavity of a sphere equal to the lunar orbit. But this orbit having its diameter about 224 times that of the moon, and the surface of a sphere being equal to four of its great circles; the secondary light which would reach the earth must

must be attenuated not less than *two hundred thousand times*, for $4\,(224)^2 = 200{,}704$. Such perfect reflection, however, cannot be admitted. If we examine the face of the moon with a good telescope, we discern round spots of extraordinary brightness, and perceive large spaces which are remarkably obscure. It is evident then, that but a very small part of the incident light must be reflected, the rest being absorbed. The quantity of reflection from paper, plaster, and other white rough surfaces, according to Bouguer himself, constitutes only the 150th part of the whole incidence. If the exterior crust of the moon resembled, therefore, any earthy body with which we are acquainted, her pale borrowed light would be at least one hundred times feebler than is actually observed. Hence I am disposed to think, that the rays of the moon are principally, if not entirely, discharged from her own mass, and that the lunar surface is of a nature analogous to the carbonate of barytes and other phosphorescent substances, which, after a partial calcination, are capable of being excited by the action of the solar rays to disengage their latent light.*

* See Note XLIII.

But since light is not homogeneous, we may presume that its several component rays have also different calorific powers. Nor can those powers be inferred from its aptitude for the purposes of vision. It is the depth, rather than the distincness, of the impression made upon the retina, that seems to mark the energy of the incident rays. Among the various colours, there is a wide diversity of character. We are all sensible of the feebleness of blue, and the softness of green; and the eye is very soon fatigued, by the dazzling glare of scarlet. But the photometer exhibits a similar gradation of force in the solar spectrum, where the primary tints and shades are developed. At the extremity of the violet, the effect is scarcely perceptible; at the termination of the red, it appears the greatest; and the several intermediate colours constitute a progressive force, nearly as the square of their distance. Thus, the numbers 1, 4, 9, and 16, respectively denote, with tolerable exactness, the calorific energies of the blue, the green, the yellow, and the red. And hence the red rays are three times more powerful than white or compound light.

Such

Such are the comparative measures of the spectrum formed by a prism of flint-glass. With one of another sort of material, though the results might bear a kindred resemblance, they would essentially differ in their proportions. Each ray suffers a refraction which is determined only by its peculiar relation to the diaphanous substance. There is no common affection that pervades them all; and therefore the several coloured pencils are, according to the nature of the prism, liable to be variously dilated. Nor does it seem possible either to assign the number, or to distinguish with certainty the limits, of the spaces that compose the spectrum; for they appear to melt into each other by a series of imperceptible gradations. The celebrated enumeration of *seven* primitive colours, and their fanciful analogy to the divisions of the diatonic scale of music, betray the spirit of mysticism which infected the seventeenth century, and which seems not yet entirely banished.*

It is hard to determine what degree of influence can be strictly ascribed to *colour*. Those different impressions most probably are caused merely by

* See Note XLIV.

the various density of the pencilled spaces. But how shall we decide, that the violet rays are con stantly more diffuse than the red? Such a conclusion, however, is most agreeable to the general phænomena, and best accords with that simplicity which is the aim of all our researches. Colour may consist in the different velocity, and therefore different impulsive force, of the rays. The eye can distinguish not only the quantity, but the quality of impression made upon the retina; that is, the species of tint as well as the degree of brightness.—It is only near the limits of contact that elementary corpuscles display their specific qualities; at remoter distances, they gradually assume a common character. Though the particles of light, according to the various colours which they are fitted to excite, suffer, in their passage through diaphanous substances, a certain diversity of attraction; yet they may exert on each other an equality of repulsive force, when, disparted at wide intervals, they become latent and constitute heat. The multifarious composition of light is, therefore, entirely compatible with the homogeneous arrangement and constitution of heat.

To

To perform judiciously those experiments with the prism, requires nice attention. It is always difficult to define the termination of the spectrum, either above or below, but especially at the edge of the red, which shades off into a dusky brown. But such indistinctness is farther augmented by another cause, that may prove a source of notable error. Besides the direct rays of the sun, the prism transmits a quantity of collateral or adventitious light, which is copiously deflected and sent from the adjacent portion of the sky. This extraneous accession forms, at both ends of the spectrum, an extending appendix of mixed or white light, which invades the obscure boundaries of colour. The degree of secondary illumination so produced is, in our foggy climate, often not inconsiderable; and in certain positions of the prism, it is collected on a narrow spot. When the tints of the spectrum are well unfolded, the reflex rays sent from a space of ten degrees in the muddy atmosphere, are sometimes all concentrated into a pencil of about one degree. But a small quantity of light, after being thus condensed, will make a very sensible impression. Under such circumstances,

circumstances, therefore, the calorific effect may become perceptible, nay powerful, even beyond the apparent limit of the red. Towards the other extremity, however, of the spectrum, this disturbing influence will be dilute and evanescent: it will then have only some tendency to augment the action of the orange, the yellow, or the green; for its force must evidently decline on either side of the red, which is the centre of energy.*

We may now perceive distinctly the reason, why the calorific impression of light was much greater upon tin-foil than upon gold-leaf. The former, being of a livid blue, must reflect the feeble rays, and, of course, absorb the more powerful. The gold surface, on the contrary, having its colour of a reddish yellow, will reject the admission of those very rays which are, from their nature or appropriate condensation, the most energetic. By covering the balls of the photometer with leaves of different metals, the various absorbent qualities corresponding to their several tints might easily be discovered and compared. Unfortunately those metallic tints are never permanent, but

* See Note XLV.

change

change their depth and character with the progress of oxydation.

A similar method of investigation may be employed for ascertaining the disposition of other coloured substances with respect to light. By painting, for instance, the dark ball of a photometer successively with the different water-colours, we shall, from their varied effects, deduce the corresponding powers of absorption. They form this series: white—red—yellow—green—blue—black. I could easily enlarge the enumeration, if its subdivisions were more constant and regular; but the disposition to absorb the light depends not more on the sort or quality of the tint, than on the depth or intensity of shade. It would be superfluous, therefore, to state the relative proportions.

Opaque coloured bodies must evidently reflect all the rays which they do not absorb The one property increases as the other diminishes; and their united force equals the entire energy of the incident rays. But these reflective powers are easily discovered, by extending sheets of stained paper in succession behind the instrument, and

screening

screening it from the direct light of the sun. In this research I had made considerable progress, till, after weighing the subject maturely, I resolved to desist. In fact, the results are merely individual, and furnish no certain general conclusion. If an extensive scale of colours were composed on the plan of Mayer or Lambert, we might then proceed with confidence. The language of description would be fixed and determinate.

But the principle on which the photometer is constructed, admits of being somewhat diversified in its application. Instead of receiving the direct impression of light, the sentient ball may be affected merely by a derivative energy. If a flat plate be exposed to the action of perpendicular rays, it will imbibe and retain twice as much heat as a globe of similar colour and materials; for the latter, with double the measure of surface on either side, intercepts only the same quantity of incidence, as does its section or its generating circle. Could we procure a metal capable of absorbing the whole of the incident light, it would experience even quadruple the ordinary effect, since

since, by its constitution, it is disposed to cool twice as slow as glass. If a metallic plate has its anterior surface japanned or blackened, it will acquire an intermediate elevation of temperature. The balancing power, or its rate of cooling, compared with that of black enamel, is evidently as $1\frac{1}{2}$ to 2; and therefore the degree of effect which it receives has to what is experienced by a ball of this material, the ratio of 8 to 3. That effect, however, is not directly measurable; but if one of the balls of a differential thermometer be placed contiguous or very near to the recipient surface, it will thence receive a proportional impression. Without stopping to investigate strictly this reflex influence, we may easily discover an approximation quite sufficient for our purpose. A vitreous or painted surface, it was observed, disperses half its heat by pulsation; but an equal surface of the same kind, placed adjacent and parallel to it, will intercept the whole of those vibratory energies. Every portion of this, except towards the extreme edge, must also be alike affected. On a glass ball however, such impression will be somewhat modified: the nearer hemisphere may feel the entire action, but the

the farther one will, from its obliquity, receive scarcely the half. Hence the joint calorific effect produced on the ball, being equal to $\frac{4}{3}(\frac{1}{2}+\frac{1}{4})=1$, is the same as what would have resulted from a direct excitement of the rays. If this approximate ball is likewise black, the combined impression will now be greatly increased; yet not quite doubled, since the light is intercepted which should have acted on that part of the surface immediately behind it. We may therefore estimate the whole accumulated energy at about $1\frac{3}{4}$; nor though the ball should absolutely touch the dark absorbent surface, would this circumstance occasion any sensible augmentation, the contact being evidently confined to a very narrow spot.

It is easy to reduce these ideas into practice. Let a differential thermometer be prepared, without any cylindrical reservoir, and without having its balls bent. The tube to which the clear ball is blown should be of uniform calibre, and a full inch shorter than the other. A thin plate of tin, brass, or silver, of about $2\frac{1}{2}$ or 3 inches in diameter, and with a hole drilled through its centre sufficient to admit the black ball, has its upper side blackened

blackened with China ink, or with ivory-black dissolved in alcohol. This plate being passed over the ball, the hole is filled up by means of two semicircular bits embracing the tube: the instrument is next inverted, and the plate, resting against the ball, is secured in that horizontal position, by means of a narrow ring of cement applied above it. The whole is then inclosed within a compound glass case, of which the upper portion is the larger segment of a globe of 3½ or 4 inches in diameter.

Another mode is, to give the differential thermometer its original form; both the balls being transparent, and of the same height, but separate from each other about 2½ inches. It is fixed into a thin wooden box, of a square or oblong shape: a tin plate near three inches broad, and of an equal length with the box, or two inches longer than the instrument, with its anterior surface blackened, is made to slide immediately behind the black ball: about an inch before it, the frame receives a pane of choice glass; and the opposite side of the box consists of a parallel plate of polished metal, at an interval of three quarters of an inch from the absorbent surface.

To

To measure the diffuse light of the sky, the box is set horizontal: but to compare that of the sun, especially during the winter season, it is placed in a vertical position. The latter method is best adapted for estimating the absorbent powers which belong to different colours. For that purpose, the metallic slide, instead of being blackened, is painted with the successive shades. And to compare more nicely the kindred or proximate tints, another similar and collateral slide is introduced.— I will not enlarge on this topic, but I would observe, that the results must be inaccurate, unless the light falls nearly perpendicular on the coloured surface. Of oblique rays, the action is proportioned to the sines of their angles of incidence. Hence this construction of the photometer is not fitted for general use.

But the sensibility of the photometer can be very considerably augmented by help of a judicious combination of cases. If the black ball be encircled by a series of concentric shells of glass, though they freely admit the influx of light, they will yet greatly retard its subsequent dispersion in the form of heat, and therefore occasion a high degree

degree of accumulation. Nor is the impression thence excited at all disturbed or diminished by any counteracting efforts of the clear ball, which, being situate without the enclosure and in open space, maintains the temperature of the atmosphere. The manner of disposing and adapting the several parts of the instrument is best understood by inspecting the figure (See fig. 37). The spherical shells are chosen as thin and clear as possible, their diameters rising in regular succession, by a difference of at least half an inch. They are each composed of two segments, cut from the globes usually blown for watch-glasses; and it is not a difficult matter to match them properly. Before joining the tube which is terminated by the black ball, the smaller segments, having a little orifice and neck, are slipped upon it. It this state, the differential thermometer is constructed; nor till after it is quite adjusted, are the shells completed. The smaller segments are first cemented to the tube at due intervals, and to these are next joined the larger ones, by means of isinglass. To strengthen the central tube, it is inclosed by a narrow pillar, consisting of two opposite pieces; and the whole is rendered more secure, by a broad piece of wood which connects this pillar with the scale attached to the other branch. For the sake of the symmetry of appearance, the top of the pillar may be formed into a bead of the same size and altitude as the clear ball. The diameter of the central ball and its successive encircling shells, as represented in the figure, are respectively 3-4ths of an inch, $2\frac{1}{2}$, $3\frac{3}{4}$, and five inches. Their surfaces are as 1, 11, 25, and 44; and hence the influence which they exert to impede the consumption of heat and thus augment the calorific effect, is easily computed. The power of one case is denoted by $\frac{12}{11}$, that of two cases by $\frac{3}{2}$, and that of all the three cases by $\frac{11}{5}$. This progression seems very slow; owing partly to the vitreous quality of the surfaces, and partly to their great inequality. It might be preferable to widen the vacant space around the black ball, and, instead of concentric shells, to employ flattish segments of large spheres; the upper pieces only being of glass, and the under ones consisting of polished metal By each of these additions, the impression made on the instrument would be increased

creased by more than one half. Six cases might therefore increase tenfold the sensibility of the photometer.

Such peculiar delicacy, however, can be procured, at least for perpendicular rays, more easily and effectually by another mode of composition. Figure 38 represents this new arrangement. The upper shells are the smaller segments of concentric spheres; the under ones consist of tin or very thin brass, which, though likewise concave, are rather flatter. The central ball is black, and the broad hollow surface which it touches is painted with China ink. The diameter of that ball is 8-10ths of an inch, and the diameters of its several cases are 4, 5.6, and 7.2 inches. The parallel surfaces are consequently as 25, 49, and 63; and hence we deduce the augmented effect. If those surfaces had been all vitreous, the increased sensibility of the instrument would be $= \frac{9}{4} \times \frac{7}{4} = \frac{63}{16}$; but if they were all metallic, it would be denoted by $\frac{16}{5} \times \frac{7}{4} = \frac{28}{5}$. Assuming therefore the mean, we may reckon the impression as heightened, by that combination, nearly five times.

I scarcely need observe that, its sensibility being thus so much increased, the instrument must of

 necessity

necessity have a shorter range. It is then fitted only for measuring the weaker lights. Those forms of construction, too, where the clear ball is extruded, are liable to some aberration from the action of wind. But this error, which proceeds merely from the variable relative temperature of the external case, is greatly diminished in consequence of the combination of surfaces, and might be avoided altogether, by inclosing the whole apparatus within a glass receiver.

These susceptible though complex photometers are therefore calculated to measure the most delicate shadings of light. It is not merely an object of curiosity, to mark the progress of the morning's dawn; an accurate solution of the question might furnish *data* for ascertaining the height and the constitution of our atmosphere. From the duration of twilight there has indeed been derived a specious, if not probable, estimate of the limits of that subtle fluid which encompasses the terraqueous globe. But Lambert has investigated geometrically the succession of crepuscles, and has proved, besides the primary one, the existence of a second, a third, and

and even a fourth, reflexion from the sky. The former conclusion is thence rendered very doubtful; and both for this and other reasons, I am inclined to believe, that our atmosphere has far less altitude than is commonly supposed. A correct series of photometrical observations could not fail to lead to a sure decision. The station for making such a register would be some convenient spot within the polar circle, where continued twilight lasts during a great part of the year, and where, for weeks and months together, the sun totally disappears. The clearness and serenity of that climate, during the sombre passage of winter, would also prove highly favourable. Repeating the observations at different hours of the day, and classing all of them according to the calculated depression of the sun below the horizon, we might assume only the mean results. An astonishing accuracy could be thus attained.

There is yet another photometrical combination, but which is adapted only for the mensuration of parallel rays. It is a small reflector having the black ball of a differential thermometer fixed in its focus. For the reflector, I prefer plate-metal, or copper coated with silver, and which is nicely hammered into a parabolic figure, its depth being rather more than the fourth part of its diameter. The differential thermometer consists of straight parallel branches, about an inch and half separate. The black ball being adjusted to the focus of the reflector, its tube is let into a small hole filed at the lower edge and firmly cemented, the clear ball standing directly in front. A convex circle of glass is then fitted into the ledge of the reflector, and this compound apparatus is finally soldered at the back to a solid pillar. Thus arranged and disposed, the instrument will indicate with precision the variable quantity of light which penetrates into an apartment. For that purpose, we have only to place it at the remote end of the room and directly facing a window.

But to produce intense effect, the reflector may be constructed of such large dimensions as to include the differential thermometer within it. By this plan, the calorific power of the moon's rays could at last be rendered visible. It might require a speculum three feet in diameter and ten inches deep; the differential thermometer being fastened

fastened in its place to two parallel wires stretched across through the plane of the focus, and the whole covered by a circular sheet of thin plate glass neatly implanted. Suspending the apparatus on swivels, it may be presented and kept steady in any oblique position; and the corresponding alteration of the zero of the scale can then be observed, and the variation of the magnitude of the degrees easily computed. Supposing the reflection to be perfect, and the black ball to have three quarters of an inch in diameter, the impression would be augmented 2204 times. Therefore, after making every reasonable allowance for the defect of performance, the force of the moon's light would excite an effect of more than a degree, and consequently would be quite perceptible. May I presume to hope that this magnificent experiment will be executed at some no very distant period?

## CHAPTER XXI.

THE operation of the photometer, we have seen, depends on the mutual balance of the action of light with its subsequent effort to diffuse itself in a latent form. The calorific effect is thus determined, not less by the measure of original excitement, than by the cumulative accessory power of dispersion. That excitement is the joint result of the quantity of incident light, and the absorbent quality of the receptive ball; but its intensity of impression must be reciprocally proportioned to the energy of the refrigerating process. With an equal degree of absorption, a metallic surface is twice as much affected as one of glass, because by its constitution it cools twice as slow. But where the force of the rays and the quality of the receptive surface remain the same, the actual change of temperature which it experiences will yet depend on the nature of the surrounding medium. If this fluid conducts away the heat more slowly than before, the impression produced

produced will of course be augmented; or if, on the contrary, it makes the transference with more rapidity, the effect will be proportionally depressed.

Hence the photometer is fitted to discover with delicate precision the relative conducting powers, not only of different gases, but of the same gas in its various states of modification. For that purpose, I used a differential thermometer of the simplest construction, only having its sentient ball formed of black glass: the branches were near eight inches long, and about an inch and half separate. A strong tube, almost two inches wide and six inches in length, was ground square at each end, and the upper one cemented to a glass plate, which had a large round hole cut through it, nearly corresponding to the diameter of the tube. Into this cavity, the differential thermometer was let down, till the balls stood about an inch and three quarters above the rim, and there secured with soft cement; a syphon mercurial-gage also being introduced and fixed beside it. The other end of the tube or pillar was next cemented to the transferer of an air-pump; and

a ball

a ball of clear glass, of four inches in diameter, and carefully ground at the bottom to an aperture of about two inches and a half, was occasionally planted on the rim or plate, the contact being, as usual, rendered air-tight by a slight application of hog's lard. By screwing the transferer to the pump, I could easily extract the air which encircled the instrument, and then turning the cock and unscrewing the apparatus, I transported it in that state to the garden. Opening the cock gently, I could again admit the air at pleasure, and could regulate the degree of rarefaction by help of the included gage.—This apparatus was equally convenient for observing the conducting powers of other gaseous fluids. The gas submitted to examination was collected under a very large receiver plunged in a water bath, and communicating with the air-pump by means of a long copper tube furnished with a stop-cock. Exhausting the cavity of the globe and its annexed cylinder, the void was immediately supplied by opening a communication with the reservoir. When very great accuracy was required, I repeated this operation two or three times, that the

the residuum of atmospheric air might be expelled or at least extremely attenuated.

We should naturally presume, that the rarefaction of the air must diminish its power of conveying heat. This conjecture is completely established by observation; but it is a point of more difficult research, to assign the law of such decrease. I soon remarked that, at each repetition of a certain number of strokes of the air-pump, the photometer successively rose by nearly equal ascents. But the corresponding rarity of the air must evidently have increased in a geometrical progression. Distinctly to ascertain, therefore, the coincidence of the supposed law or its limits of deviation, I proceeded by a continued series of bisections, or at least quadrisections. For making these experiments I chose the finest weather, when the sun was bright and the sky unclouded. The operations were generally performed during the space of an hour before and after mid-day; in which interval, the rays suffered scarcely any visible alteration of force. But guided by the contemporaneous indications of another photometer, I was able to reduce all the

the calorific effects to the same standard. It will be sufficient, therefore, to state the proportional mean results, which were derived from the comparison of numerous observations.

## EXPERIMENT LIV.

I rarefied the air contained within the compound photometer successively 4, 16, 64, and finally 256 times; and found the corresponding impression of the sun's rays to mount, from 100 degrees, to 120, 141, 162, and at last to 185.—

These quantities form a series evidently akin to the geometrical; for their differences continually increase, though with a very slow progression. They are inversely proportioned to the disposition of air, when differently rarefied, to abstract heat from a vitreous surface. It would, however, be rash thence to infer in general the precise influence of rarefaction in diminishing the conducting power of air. The discharge of heat, being composed of distinct elements, may therefore be variously affected by the same causes. Thus, rarefaction may perhaps occasion a different modification

cation of the pulsatory energy, from that which it produces on the abductive or the radical part of communication. And this surmise was actually confirmed by experiment. On covering the black ball of the differential thermometer with a bit of tinfoil, I found that the same progressive rarefaction now caused a rapid and accelerating augmentation of effect. This invigorated action is rendered still more conspicuous, by opposing glass to metal. Having transfered the tinfoil to the clear ball, I remarked that the coloured liquor subsided a little at first, but soon rose again; and as the air became gradually rarer, it pushed upwards with increasing celerity. It is obvious, that only the difference between the refrigerating power of glass and metal in contact with rarefied air, was here exhibited. The seeming anomaly was occasioned by the copious absorption of the black vitreous surface; but this preponderance was quickly more than counterbalanced by the rapid accumulation of energy on the surface of the metal, in consequence of its diminished power of dispersion. Hence, even when the black ball was covered with tin, there must have

existed

existed a certain degree of irregularity, owing to the influence of the clear ball, which, always absorbing some light however small, would yet be affected in a different proportion. To obtain correct results, therefore, it is necessary that each ball should present a metallic surface. With this view, I constructed another more sensible differential thermometer, and had its balls, which were both clear, doubly gilt, the one with gold and the other with silver. The yellow surface, of course, imbibed more light than the white, and the difference of the proportional effects was marked by the instrument.

## EXPERIMENT LV.

Having, as before, rarefied the contained air successively 4, 16, 64, and at last 256 times, I found, that the corresponding impression made by the sun's rays on a bright metallic surface rose progressively, from 200 divisions, to 270, 362, 477, and finally to 620.—

Since, in the case where only small differences of temperature are concerned, a metallic surface cools

cools twice as fast as a vitreous one,—these numbers, which commence in that proportion, must express photometric degrees. They increase however with a much swifter progression, for their last term is more than triple that of the former set of quantities. They approach evidently to the nature of a geometrical series, yet their acceleration is not quite so rapid; the middle term being 362, while the mean proportional between the extremes, or 200 and 620, is only 352½.

But to discover the true principle of relation, we must analyse the refrigerating process. Let the discharge of heat from a vitreous surface immersed in air of the ordinary density, be denoted by 100; then the measures of dispersion corresponding to the progressive scale of rarefaction, and which are reciprocally as the impressions made on the instrument, will be represented by the successive numbers, 100, 83, 71, 61, and 54. By a similar mode of procedure, we find the dispersive flow of heat from a metallic surface, corresponding to the same range of dilated air, is expressed by 50, 37, 27, 21, and 16. The respective differences between this and the preceding

ing set of numbers, are 50, 46, 44, 40, 38; which form a new progression, representing the excesses of the pulsatory energies of glass above those of metal. If the relative proportions, therefore, of such energies be not affected by the rarity of the contiguous air, those several terms, augmented by one-seventh part, will denote the whole of that species of expenditure which belongs to a vitreous surface. Hence, corresponding to the same scale of rarefaction, the series 57, 33, 50, 46, and 43, represents the portion of heat discharged from glass by pulsation; and the supplementary series, 43, 30, 21, 15, and 11, exhibits the portion dispersed by abduction, and which is the same for every kind of surface.

This last series is clearly geometrical, every second term of it being derived from a bisection; thus 43, 21, 11, and 30, 15. The other series, which expresses the force of pulsation, is apparently of the same kind, but declines at a very slow rate. The difference between the last term is almost three-fourths of the first; and the square of $\frac{3}{4}$ is $\frac{9}{16}$, or about one half. A bisection would, therefore, occur at twice that interval, or at every tenth

tenth term. But the initial terms, 43 and 57, of both ranges are as 3 to 4, which the conditions of the case indeed required; for the compound discharge from glass must be the double of that from metal, or $3 + 4 = 2\ (3 + \frac{4}{8})$. Hence, by collecting these facts together, we learn that, if D represents the density of the contiguous air, the discharge of heat from a vitreous surface will always be denoted by $\frac{1}{7}\ (3D^{\frac{1}{4}} + 4D^{\frac{1}{20}})$, and the discharge from a metallic surface will be expressed by $\frac{1}{7}\ (3\,D^{\frac{1}{4}} + \frac{1}{2}\,D^{\frac{1}{20}})$. These formulæ are perfectly consistent, and nowise intricate. It might have been desirable to have extended the ranges from which they were deduced; but though my air-pump is one of the best and most expensive sort, I could not with certainty push the rarefaction much farther.

But whatever has the property of dilating the air, has likewise a tendency to depress its power of abstracting heat. The various elastic vapours, in mixing with the atmospheric fluid, communicate their peculiar expansive force, and seem by that union to occasion a proportional diminution of its conducting quality. The influence of humidity is visible, but the more volatile substances have a marked effect. Having introduced, within the photometric apparatus, a bit of flannel thoroughly dried, the impression of the solar rays was 100°; but removing this and damping the contained air, the liquor appeared to drop near one degree: on pouring into the instrument some alcohol, it fell to 102°; and afterwards, on throwing down a small portion of sulphuric æther, it sank to 105°. By rarefaction, these effects are proportionally much augmented. Thus, air reduced to the fourth of its usual density, and which, under the same standard, would mark an impression of 120°,—indicates 130, when included with alcohol,—and not less than 150 or 160, if charged with the vapour of æther. I state these measures for the sake of illustration, and not as correct results. In fact they are subject to great variation, from the influence of external heat and other modifying causes. There is besides a circumstance deserving notice, and which may sometimes occasion considerable error. When the humidity or vapour is profusely generated, it collects and condenses near the top of the glass cap, covering

covering the surface with minute globules. The direct rays have thus their force somewhat impaired; but the absolute quantity of light received on the black ball is yet increased, as in the case of snow, by the copious reflection from such a multitude of glistening points. The augmentation of effect arising from this source would, I found, amount fully to the tenth part of the whole.

The permanent gases differ as much from common air, perhaps, by their disposition to conduct heat, as by their density or other properties. The azotic and the oxygenous, indeed, seem to possess it nearly in the same degree. But carbonic gas abstracts the heat from a vitreous surface about an eighth part slower, and from a surface of metal one-fourth slower, than common air. By progressive rarefaction, that property is also reduced on a similar scale. Hydrogenous gas, however, is the most distinguished by its affection for heat, which it conducts with unusual energy. And as it is so easily procured in large quantities and in a state of tolerable purity, my observations were principally directed towards that singular fluid.

## EXPERIMENT LVI.

Having filled the compound photometer with hydrogenous gas and exposed it to the direct rays of the sun, I found that, reckoning their force as before equal to 100, the impression made upon the black ball was only 44 degrees, and that upon the gilt one 56.—

Thus the contact of hydrogenous gas does more than double the expenditure of heat from a vitreous surface, and accelerate the process of refrigeration almost four times from a surface of metal. This inequality of effect proves its influence to be exerted chiefly, if not entirely, in augmenting the abductive portion. The reciprocals of those quantities, or 228 and 179, must express the measures of communication which respectively belong to a vitreous and a metallic surface. Their difference is 49, and consequently 7 denotes the pulsatory energy of metal, and 56 that of glass. But this result is obviously the same as what was obtained in the case of atmospheric air. The remainder, or 172, = 228 — 56 = 179 — 7, is equal to four times 43; that is, the hydrogenous gas, without

without altering the force of pulsation, has quadruple efficacy in abstracting heat by the process of abduction.

## EXPERIMENT LVII.

I filled the apparatus with hydrogenous gas, which I successively rarefied, 4, 16, 64, and finally 264 times: the corresponding impressions were 44, 54, 66, 80, and 96 upon the black ball, and 56, 73, 95, 124, and 160 upon the gilt balls.—

Of the first series, the reciprocals are 228, 184, 150, 125, and 104; and of the second series, they are 179, 137, 105, 81, and 62. Their mutual differences, being increased as before by the seventh part, give the progression 56, 54, 52, 50, 48; which represents the pulsatory energies of a vitreous surface. The remainders, 172, 130, 98, 75, and 42, exhibit the abductive powers of any surface immersed in hydrogenous gas. These numbers are evidently in a continued ratio, the last term being only the fourth part of the first. The former series, though apparently arithmetical, we may conclude from analogy to belong likewise to the geometrical kind: on examination, we shall discover that, at every fifteenth term a bisection would take place. Hence are derived the formulæ to denote the refrigerating power of rarefied hydrogenous gas: for a vitreous surface it is $\frac{1}{7}(12 D^{\frac{1}{5}} + 4 D^{\frac{1}{30}})$, and for a metallic surface it is $\frac{1}{7}(12 D^{\frac{1}{5}} + \frac{1}{2} D^{\frac{1}{30}})$.

It may now be eligible to exhibit in a collective view the component elements of the expenditure of heat from a vitreous and a metallic surface, immersed either in common air or in hydrogenous gas of various densities. This table will farther elucidate the theory which we have been developing, and will furnish matter for curious and interesting speculation.—

| | Atmospheric Air | | | | | Hydrogenous Gas | | | | |
|---|---|---|---|---|---|---|---|---|---|---|
| | | Vitreous Surface | | Metallic Surface | | | Vitreous Surface | | Metallic Surface | |
| Rarity | Abductive Power | Pulsatory Energy | Expenditure of Heat | Pulsatory Energy | Expenditure of Heat | Abductive Power | Pulsatory Energy | Expenditure of Heat | Pulsatory Energy | Expenditure of Heat |
| 1 | .4286 | .5714 | 1.0000 | .0714 | .5000 | 1.7143 | .5714 | 2.2857 | .0714 | 1.7857 |
| 2 | .3604 | .5519 | .9123 | .0690 | .4294 | 1.4924 | .5584 | 2.0508 | .0698 | 1.5622 |
| 4 | .3030 | .5332 | .8362 | .0667 | .3697 | 1.2993 | .5456 | 1.8449 | .0682 | 1.3675 |
| 8 | .2548 | .5150 | .7698 | .0644 | .3192 | 1.1311 | .5331 | 1.6642 | .0666 | 1.1977 |
| 16 | .2143 | .4975 | .7118 | .0622 | .2765 | .9847 | .5210 | 1.5057 | .0651 | 1.0498 |
| 32 | .1802 | .4805 | .6607 | .0601 | .2403 | .8571 | .5091 | 1.3662 | .0637 | .9208 |
| 64 | .1516 | .4641 | .6157 | .0580 | .2096 | .7462 | .4974 | 1.2436 | .0622 | .8084 |
| 128 | .1274 | .4483 | .5757 | .0560 | .1835 | .6496 | .4861 | 1.1357 | .0608 | .7104 |
| 256 | .1071 | .4331 | .5402 | 0542 | .1613 | .5655 | .4750 | 1.0405 | .0594 | .6249 |
| 512 | .0901 | .4183 | 5084 | .0523 | .1424 | .4824 | .4641 | .9565 | .0580 | .5504 |
| 1024 | .0758 | .4041 | 4798 | .0505 | .1263 | .4286 | .4535 | .8821 | .0567 | .4853 |

The same mode of computation, it is evident, will extend equally to the case of repeated condensations. Those results might likewise be verified by experiment; but such experiments are of most arduous execution, and unfortunately confined to a very narrow range. It is difficult to condense the air four times, to condense it more than sixteen times is hardly practicable; and even then, it would require uncommon precaution, and a thickness of glass most unfavourable to the admission of light. The conducting quality of air having sixteen times the ordinary density is only 1.5136 from a vitreous surface.

If the air included within the case of the photometer communicates, however imperfectly, with the external atmosphere, it must suffer the same variation of density, produced by the change of temperature and the fluctuation of barometric pressure. Hence the instrument is subject to a slight modification in the scale of its measures. With the same intensity of light, it will indicate a greater impression, if the air grows either lighter or warmer. For each degree centigrade of alteration of temperature, that difference will be the

263d part; and corresponding to every tenth of an inch in the height of the barometer, the variation will only be equal to the 316th part. Such corrections may, therefore, in general be disregarded. Indeed they can very seldom cause an aberration amounting to the twentieth part of the whole; and were the joining of the case perfectly tight, they would be rendered unnecessary altogether. But if the change of density be very considerable, air will slowly transpire through the pores of the wood. On lofty summits, therefore, it would be preferable perhaps to permit such communication and make the proper allowance. The density of the air is .575 on the top of Mont Blanc, it being as unit at the level of the sea. Consequently its conducting power, at such an elevation, is .929, or it is diminished by the fourteenth part. The force of the sun's rays, in that thin atmosphere, is, with 60° of incidence, about 1.15, compared with what would obtain under the same obliquity in the plain below. I should thence expect the whole impression to be 1.24.

The conclusions respecting the elements of the refrigerating process as affected by the nature of the

the ambient medium, which have thus been derived from photometric observations, are satisfactorily confirmed by direct experiments performed on heated substances themselves, within receivers either partially exhausted or filled with different gases. These experiments, however, are attended with considerable difficulties. Water, which proved so convenient in our former researches, is here precluded: we must employ some fixed liquid, such as oil; which is not only of troublesome management, but is little capacious of heat. I was obliged to accommodate the apparatus to the size and form of the receivers in my possession; and the whole of the operations were conducted on a comparatively miniature scale.

I procured a slender mahogany frame, consisting of two thin circular pieces connected horizontally by four delicate pillars, standing upright or rather with a small convergency. (See Pl. IX.) The upper piece was only a ring of 4½ inches in diameter and half an inch broad: the under one was five inches in diameter, and had a hole scooped out of it to receive an elliptical reflector of plate metal, 4½ inches in diameter and 1¼ deep. The differential thermometer

thermometer was proportionally small, being of the shape of the letter V, and placed in a horizontal position; it was cemented at its angle to one of the pillars, and its remote ball lay between the receiver and the reflector, the wooden rim being there filed away to enlarge the vacant space. The transverse pieces were seven inches asunder, and the pillars extended downwards an inch and half farther. This compound apparatus, was for greater convenience, set upon the plate of the transferer, now screwed to the air-pump. A circular canister 5 inches broad and 1¾ deep, stood on the upper ring, and being filled with oil heated to 100° or 150°, a fine thermometer was inserted, resting against the bottom. To prevent the vapour of the oil from condensing on the receiver or reflector, I lapped round the upper part of the pillars a ribband of flannel thoroughly dried, and which, therefore, absorbed the moisture as fast as it exhaled. The glass receiver was of a cylindrical shape, about 12 inches high and 6 wide, fitted as usual to the plate by means of hog's lard. I then worked the pump vigorously; but before the exhaustion had proceeded far, the heat was greatly reduced, and it seldom

seldom amounted to 50° when that steady equilibrium took place which is required for accurate observation. I shall here only state the proportional results deduced from repeated trials.

## EXPERIMENT LVIII.

I covered the bottom of the canister with thick bibulous paper soaked in oil, and disposed the apparatus as usual for action. The effect produced on the focal ball was 100°; but after rarefying the air within the receiver 64 times, it rose to 132°. Having refilled the canister with hot oil and exhausted the receiver, I now admitted hydrogenous gas; the impression was only 44°: it mounted however to 70°, on rarefying that gas 64 times.—The paper being removed from the bottom of the canister, effects exactly proportional, though much smaller, were excited, and which ranged, according to the degree of polish and metallic lustre, between the third and the sixth of the former measures.—

Thus the force of pulsation is very sensibly augmented by rarefaction. But we discover on reflection,

flection, that this increase is only apparent, and that the pulsatory energy is then really diminished. Since the consumption of heat from a vitreous surface in air rarefied 64 times, is denoted by 6157; the same vigour of pulsation would have excited an impression equal to $162\frac{1}{2}°$. Hence the pulsatory energy had been actually reduced in the proportion of $162\frac{1}{2}$ to 132; which corresponds very nearly with that of .5714 to .4641, or of $\sqrt[20]{64}$ to 1, as determined before. If the abductive power had diminished by rarefaction as fast as the pulsatory energy, it is obvious that the effect on the focal ball would have continued unalterably at 100°; nor would this experiment have detected any diminution of intensity in either of those elementary processes.

In hydrogenous gas, the impression was reduced to 44°; but this is inversely as the superior power of the gas to abstract heat from a vitreous surface. Therefore, in hydrogenous gas, the pulsatory energy continues exactly the same, as in atmospheric air. It is likewise enfeebled by rarefaction, though somewhat more slowly: for the rate of cooling being depressed from 2.2857 to 1.2436, the

the original energy would have produced an impression equal to $80\frac{1}{2}°$; and hence that force is diminished in the proportion of $80\frac{1}{2}$ to 70, which corresponds to that of .5714 to .4974, or $\sqrt[30]{64}$ to 1, as before ascertained.

The principles above deduced from photometric observations, receive entire confirmation from the various rates of cooling experienced by the same body, on immersing it in gas of a different species or density. I here operated on a still smaller scale: a round tin canister, of two inches in diameter and height, was filled with oil of almonds, and placed with its inserted thermometer in the centre of a receiver of about 6 inches high and 4 inches wide, and standing on the plate of the transferer. I proceeded by progressive rarefactions, applying at first only a gentle degree of heat, and noting the thermometer at the end of every minute. The results coincided almost exactly with the proportions before established. To quote examples, therefore, I judge unnecessary; I shall only remark some striking contrasts of effect. Thus, the addition of a coat of pigment, which makes the canister cool twice as fast in air of the

ordinary

ordinary density, actually triples the comparative rate of cooling, when it is rarefied 64 times.. Such addition exerts a very slight influence however, in hydrogenous gas, not accelerating the process of refrigeration by more than the fourth part.—The reason is that, in common air, the pulsatory energy, being the least enfeebled by rarefaction, comes to constitute the major portion of the whole discharge; while, in hydrogenous gas, it formed but a comparatively small share.—If the rate with which the painted canister cools in air rarefied 256 times be carefully observed; on admitting hydrogenous gas, it will cool four times faster. But if the same experiment be repeated, after the surface of the canister is restored to its bright metallic lustre; it will now be found to cool no less than eleven times faster.

It would be interesting to extend similar observations to thin pellicles applied to the surface of the canister. I can only cite one experiment of that sort; but it is perfectly consonant with the general principles. Having rubbed the outside of the canister entirely over with a feather dipt in olive oil, I found that, in atmospheric air, it cooled

cooled about a tenth part faster; but, after this air was rarefied 64 times, it cooled four tenths faster. In hydrogenous gas, scarcely any alteration was perceived in consequence of the application of the oil; on being rarefied, however, to the same degree, the refrigerating process was accelerated by nearly three tenths. The coat of oil increased, though partially, the pulsatory energy; and of such augmentation, the effect has been already anticipated.

But experiments on the cooling of bodies immersed in gases of different kinds and differently rarefied, if continued through a wide extent of temperature, discover another element, of variable intensity, which enters into the process of refrigeration. It consists in the regressive motion, or perpendicular flow from the surface, excited in the ambient fluid, and which grows more rapid and efficacious in proportion to the degree of heat. This action, at the interval between boiling and freezing, becomes, in common air, nearly equal, we have seen, to the constant power of discharge from a vitreous surface, and double that from a surface

surface of metal. It is the source of the whole inequality remarked in the rate of cooling, which always betrays more or less a tendency to decline. By rarefaction, however, that accessory force appears to be extremely diminished. The canister, whether bright or painted, was found to cool with surprising uniformity in rarefied air. Before that ambient fluid, was rarefied 32 times, the distinctive quality of surface had become almost evanescent; in the whole descent of 100 degrees, the rate of cooling from a boundary of tin suffered then a retardation of only one-sixth,— and that from a coat of pigment, not more than the fourteenth part.

In hydrogenous gas, the power of recession is comparatively greater than in common air. At the elevation of the boiling point, it more than doubles the discharge from a surface of paint, and nearly triples that from one of metal. That progressive energy seems also to be less affected by rarefaction. Corresponding to the same interval of temperature, the rate of cooling from a vitreous surface plunged in hydrogenous gas rarefied 32 times,

times, is yet increased by more than the half, and that from a metallic surface, in like circumstances, receives an augmentation of above a third.

The different influence of rarefaction is best perceived, however, by comparing the numerical relations. In atmospheric air of the ordinary density, the portions of heat discharged near zero from the two opposite kinds of surfaces are as 50 and 100; but, at the boiling point, they are as 100 and 150, being there augmented by a force of recession equal to 100. In air expanded 32 times, those dispersions are, near the limit of equilibrium, represented by 24 and 66; and at the excess of 100 degrees of temperature, they are denoted by 29 and 71, having thus received only an increase of 5.—Perhaps it might be possible to penetrate the reason of such a material change of effect. The celerity of regressive flow is not altered by rarefaction; for if the dilating force be diminished, its space of action is proportionally increased. But rarefaction augments the attractive union of heat to air, which, though 32 times

 rarer,

rarer, will yet contain within the same space nearly the twentieth part of its original share. This inference agrees exactly with fact, since the measure, 100, of recession was reduced to 5.

In hydrogenous gas, the quantities of heat dispersed from a metallic and a vitreous surface are, near the limit of equilibrium, denoted by 179 and 229; but at the elevation of 100 degrees, they are represented by 479 and 529, thus acquiring an accession equal to 300. This gas, being at least nine times more elastic than common air, must have a regressive flow thrice as rapid. But it contains, under the same bulk, an equal portion of heat; and, therefore, its recession will be about three times more efficacious than before.—When hydrogenous gas is 32 times rarer, the respective discharges of heat are 92 and 137 near zero, and 142 and 187 at the boiling point. The augmentation corresponding to that interval, is consequently 50, or the sixth part of the ordinary measure. To apply the former explication, it would thus be required to suppose, that the hydrogenous gas included within the receiver, after being rarefied

fied 32 times, still contains the one-sixth of its original heat; a concession which is not easily admitted.

As rarefaction advances, the abductive power diminishes always more slowly than the regressive. Nor is it difficult to discern the probable cause. If the thickness of the stagnant shell of warm atmosphere were constant, those kindred elements of discharge would retain invariably the same mutual proportion. But as the air becomes rarer, its heated portions, then suffering less resistance, must rise upwards with redoubled celerity. Hence the limit of the stagnant atmospheric coat draws nearer the surface, and consequently the successive transfer of heat is proportionally increased. The law of contraction is not very distinctly marked: it seems, however, to follow nearly that of the velocity due to a given resistance, and therefore to approach the subduplicate ratio of the scale of rarefaction. Thus, in air rarefied 32 times, the abductive power was reduced to the twentieth part, while the regressive was diminished only $2\frac{1}{3}$ times, or was about eight

 times

times less affected. But 8 is not very different from the square-root of 32.

In hydrogenous gas, too, the same law nearly obtains. By a similar rarefaction, the abductive power was reduced to the sixth, and the regressive to the half, or was three times less affected. But if 8 exceeds the square-root of 32, 3 falls as much below it.—The coincidence is nearer on comparing hydrogenous gas with common air. In the same space, they contain equal quantities of heat; but the former, in its state of purity, being 12 times more elastic, must communicate its impressions 3½ times faster. This transfer, or the abductive discharge of heat, is actually four times swifter in hydrogenous gas than in atmospheric air.

Why the pulsatory discharge should be the same in two such different fluids, or why that energy should in general be so little affected by the progress of rarefaction—it is more arduous to explain. Hydrogenous gas transmits its vibrations at least three times faster than common air. But it must also be more than twice as capable of impression

impression from the contact with a warm surface; for the particles, in their wide distension, contain each perhaps twelve times as much heat, while only about five times fewer of them will occur in any transverse section. The salient points, or energetic particles, are consequently eight times more diffuse than in atmospheric air, or they are disparted over the surface of contact near thrice as far asunder.

The centres of pulsatory action are in every case so widely scattered, that they suffer but little derangement from the progress of rarefaction. The intervals of separation continue nearly the same, only the interjacent and inefficient particles are gradually removed. Their mutual distance, however, seems to depend in some degree on the remaining elasticity of the medium. As this diminishes by rarefaction, the salient points likewise slowly distend.—

* * * * * * * * *

# NOTES AND ILLUSTRATIONS.

## Note I. p. 3.

IT will not be judged superfluous, to describe the method which I used for striking large parabolic segments. Let AB (fig. 3) denote the extreme breadth, and CD the depth; divide AB into 20 equal parts, and draw perpendiculars from the points of section. Let CD be equal to 100 parts by any scale: make the next ordinate on either side $= 99$, or $9 \times 11$, by the same scale; the adjacent pair of ordinates $= 96$, or $8 \times 12$, and so on; those numbers being respectively as the rectangles of the segments into which CD is divided.

This procedure is founded on a very simple property of the parabola. For let the parameter, or four times the focal distance, be expressed by P; then $AC^2 = CD \times P$, and $FG^2 = GD \times P$, and consequently $AC^2 - FG^2 = CG \times P$, or $AE \times EB = EF \times P$.

By a slight alteration of the same plan, we may likewise delineate elliptical segments with sufficient exactness. For it is a general property of the conic sections, that, if two chords intersect each other, the rectangles of their segments are proportional to the squares of the parallel diameters. Hence, if the several perpendiculars EF, be augmented to E*f*, in the ratio of the semi-transverse diameter, to DG the deflection; the points, *f*, will now mark the portion of an ellipse.

## Note II. p. 5.

Let AB (fig.4) represent a small portion of a sphere, C its centre, and ACR its axis. Suppose F denotes the primary focus, or the focus of parallel rays, such as OB: if R be a radiant point, it is evident that the reflected ray B*f* will approach the radius BC, which, by the property of the circle, is perpendicular to the refringent surface at B, by the same measure of inclination that the incident ray RB approaches the vertical; for the angles FBC and *f*BC are respectively equal to OBC and RBC, and consequently their mutual differences FB*f* and OBR are equal. But OBR is equal to the alternate angle FRB, and therefore FRB is equal to FB*f*; and since the angle BFR is common, the two triangles FB*f* and FRB must be similar. Hence the analogy FR : FB :: FB : F*f*, and consequently $Ff = \frac{BF^2}{FR}$. The point F is so near A, that BF may be considered as equal to AF, and therefore $Ff = \frac{AF^2}{FR}$; that is, *the variation of the focus is directly as the square of the primary focal length, and inversely as the distance of the primary focus from the radiant point.*

Suppose the radiant point to coincide with the centre of the spherical segment, it is obvious that the rays, falling perpendicularly, will, in this case, be returned to the same point. Therefore $FC = \frac{AF^2}{FC}$, and $FC^2 = AF^2$, or $FC = AF$; that is, the primary focus bisects the radius CA.

Since $Ff = \frac{AF^2}{FR}$, the focal length A*f* must be =

$$\frac{AF^2 + AF \times FR}{FR} = \frac{AF \times AR}{FR} = \frac{AC \times AR}{2FR} = \frac{AC \times AR}{2AR - AC};$$

which last expression is the one commonly used in catoptrics.

It is plain that this mode of investigation will apply to any figure of moderate extent, since the reflecting surface may be practically considered as coinciding with a sphere of equal curvature.

## Note III. p. 11.

I shall content myself with mentioning one of the simplest and most accurate methods of obtaining the graduation of the differential thermometer. Cover the ball which terminates the naked stem with snow, pressing it all round into a compact crust, and set the instrument in a close room. As the snow gradually softens and melts away, the included ball, during that slow process, will remain constantly at zero or the point of congelation; and consequently the ascent of the coloured liquor will correspond exactly to the temperature of the other ball or that of the room, and which may be determined by a fine thermometer.

## Note IV. p. 15.

In those elementary computations, I used invariably the *sliding rule*. It is very expeditious, and I found it sufficiently accurate for my purpose. Nor can I forbear observing the unfairnesss of affecting, by the display of decimal fractions, a greater degree of accuracy than the nature of the case will admit. Calculation should never go beyond the reasonable limits of the experiments on which it rests.—I cannot help remarking by the way, that an instrument so generally useful as the sliding rule, and which was contrived early in the 17th century, soon after the beautiful invention of logarithms on which it is founded, should yet continue almost unknown upon the Continent.

## Note V. p. 27.

If a person sitting opposite to a window, gently suspends a piece of gold-leaf before his eye, he will yet perceive the external objects very distinctly, and without any distortion of figure, but tinged with a delicate greenish colour. This fact is well known, and proves decisively that the rays of light can actually permeate the substance of gold. It is evidently not through the mechanical pores or interstices of the metallic film, that those greenish rays effect their passage. On examining gold-leaf narrowly, we observe it perforated indeed with a number of minute holes, produced no doubt by the beating, and which, from the escape of white unaltered light, appear so many lucid points. But the light that is transmitted through the substance of the leaf is peculiarly modified, and must have suffered in its passage a sort of refined chemical filtration. The red and yellow rays seem to be detached from the compound beam by reflection or absorption, and the green or blue rays only are permitted to continue their course.

It may excite some surprize to find gold ranged with the diaphanous bodies: but we should recollect that, in all her productions, Nature exhibits a chain of perpetual gradation, and that the systematic divisions and limitations are entirely artificial, and designed merely to assist the memory and facilitate our conceptions. From the most pellucid to the most opaque substance, it might be possible to trace every shade of transparency. Neither glass, nor water, nor air, is perfectly diaphanous. When they are of considerable thickness, the intensity of the light which has penetrated through them, becomes visibly diminished.

If the substance which is opposed to a beam of light has its surface irregular, or its internal structure amorphous, the rays, on their emerging, will be variously turned aside and dispersed. Such is plainly the case with paper, which admits in a very sensible degree the passage of light, and yet will not enable us to distinguish the shapes of external objects. It is occasionally used instead of glass for windows, and has then the same precise effect as plates that are ground to a rough surface.

But even when, from the accuracy of the bounding planes

and the uniform constitution of the matter interposed, the rays are allowed to pursue an undeviating course, they yet suffer in their transit more or less by absorption. Nor can this absorption be reckoned only casual and indifferent; the substance penetrated, unfolding its intimate nature, exerts on the various component particles of light its specific attraction, by which certain kinds of them are detained in larger proportion than others. Thus, air intercepts preferably the blue or the green rays; and therefore white light transmitted through the atmosphere, after such defalcation, assumes, according to the length of its passage and the density of the medium, a succession of deepening tints, and passes gradually from yellow to orange, and finally dies away in a dark red. Hence the gorgeous spectacle of the setting sun, so wonderfully magnificent in the Alpine countries.

Water likewise intercepts principally the more refrangible rays; a property which seems to extend to certain solid substances of an irregular structure, such as paper or ivory. If a card be held perpendicularly against a dense pencil of white light,—for instance, against the solar rays collected in the focus of a lens, it will exhibit on its posterior surface a bright yellow circle: and if it be turned more and more obliquely, this circle will change into an ellipse with an eccentricity continually increasing, while the colour will progressively deepen into an orange, and at last a dull red. The same experiment may be performed still more satisfactorily with porcelain or white enamel.

Every one almost is acquainted with the colour of the sea, but it is not so generally known that this colour varies materially according to the depth of soundings. When the bottom consists of a white sand, the water near the shore is of a dilute green, which however grows more intense and inclines to blue, in proportion as the depth increases. The colour of the German and the Baltic seas is only a pale green, while that of the Atlantic Ocean is of a dark azure. The diversity

of

of effect is produced by the light reflected from the bottom mingling with what is sent back from the body of water. Even when the depth exceeds 50 fathoms, the reflection from the bottom will visibly dilute the radical colour. It is hence that the experienced pilot can, without employing soundings, distinguish easily an approaching sand bank.

It has been supposed, that the sea derives its greenish colour from the saline matter which it contains. Fresh water, however, in a large mass, and free from impurities, presents the same appearance. In the northern parts of Europe, this property is less observable, because our lakes are very seldom limpid, and frequently dyed with brown vegetable extracts. But in the romantic country of Switzerland, the noble collections of fresh water, being extremely clear and of prodigious depth, constantly display their natural bright green, which, contrasted with the stupendous grandeur of the surrounding scenery, has a charming effect. The rivers too, which flow from those Alpine lakes, retain, to a very considerable distance that beautiful soft colour. The Rhine, as low as Basle, still exhibits a fine green; and the turbid Rhone, after having deposited a copious sediment in its ample basin, issues forth at Geneva, with the lustre and intermediate tint of the emerald and the beryl.

I have already wandered insensibly from my subject, yet I cannot resist the inclination of using the privilege of a note, to mention here a singular phænomenon which the Swiss lakes often present, but which travellers have seldom observed, at least with attention. In certain dispositions of the sky, the green expanse of these lakes appears marked with frequent spaces of purple. Those who have witnessed that beautiful effect, have hastily satisfied themselves by attributing it to the reflection of the clouds. But it takes place frequently when no clouds disturb the serenity of the sky, or in the middle of the day, when the clouds are of a milky whiteness. It is not perceived in the mornings and evenings, when the horizon

horizon is illumed with tints of orange and red. It occurs in bright weather, when the surface of the lakes is ruffled by a gentle breeze, or mottled by the shadows of passing clouds; in short, whenever light and shade are unequally distributed. The eye is then delighted with alternate or intermingled tints of green and purple. The picture is so vivid, that we fancy it to be real; yet it is merely an illusion of sight. This curious fact belongs to the class of phænomena which authors have denominated *accidental colours*, or *ocular spectra*. If the eye be fixed on a ground of bright green, it will spontaneously fill up the intervals or shadows with purple; and conversely, if that exquisite organ be steadily directed towards a purple ground, it will almost instantly paint the vacuities, or dark spaces, with a greenish tint. To give a satisfactory explication derived from the physiological structure of the sensorium, is perhaps impossible; but the fact is easily verified. If, while the sun is shining, I hold a green umbrella expanded over my head out of doors with one hand, and in the other a piece of white paper; the paper will of course have a greenish hue, but the shadows of my fingers projected on it will seem of a purple or rose colour. And, if in the same situation, I look attentively at the leaf of a book, the characters will appear of a delicate red. In a bright day, we may remark, that, if a green curtain is dropt at the window,—while the prominences of the cornice and wainscoting of the room appear likewise green, the hollow parts appear of a pink or a purplish tint.

## Note VI. p. 28.

It is possible sometimes to determine, whether a property is inherent in the constitution of a body, or is only produced by a series of external impressions. In the former case, the effect is immediate; in the latter, it is developed gradually.

Hence

Hence gravitation is essential to matter, for its remote action is now proved to be absolutely instantaneous. The reception of that grand principle, among the learned, has been retarded by a ridiculous prejudice, as if it revived the occult qualities of the schools. Attraction is only the expression of a fact, but an ultimate fact, of which it were vain to seek an explanation. It was a capital step that Boscovich made in mechanical philosophy, when he extended the same principle to impulse and the communication of motion.

The pole of a magnetic bar attracts the dissimilar pole of another magnet, and an electrified body attracts another body of an opposite electricity: and the force thus exerted is simultaneous, and consequently inherent in the substances themselves. But if a magnet be brought near a bit of iron, or if an electrified body be made to approach another body in a quiescent state, a perceptible interval will elapse before the action is manifested. These passive adjuncts acquire, from their apposition, that peculiar internal arrangement of particles which seems to constitute magnetism or electricity, and a certain portion of time, however small, is necessary for effecting the complex motion implied in such a change of disposition.. If, instead of iron, a piece of steel be used, the effect will be still more tardily evolved, and the induced magnetism which must always precede the attractive energy, will even subsist in some degree after its cause is removed. A similar observation will apply to certain electrical phænomena; for, if the excited body be approximated to a subtance of an imperfect, or rather a slow, conducting quality, the action will appear gradually to increase.

## Note VII. p. 34.

It is well known that, water is capable of resisting for some time the process of congelation, even after it has been cooled down several degrees below the freezing point. In fact, every species

species of crystallization must require a considerable expense and duration of force, to produce that peculiar internal arrangement among the particles in which it seems to consist. But the act of freezing is farther retarded, by the time consumed in liberating the minute air globules naturally combined with the water, and which must be discharged or extricated previous to the formation of ice. In reversing the process, there is a similar expense of time. though not to the same degree. The destruction of that symmetry which constitutes the crystalline structure, requires not such nice developement of forces, and is effected with more rapidity. Yet the surface of ice, no doubt, is heated somewhat above the freezing point during the operation of thawing. This difference must be proportional to the solidity of the ice, and the warmth and activity of the surrounding air. The water which flows from it will seldom be more than a degree above the point of congelation. This water appears again to recover its portion of air by a slow absorption; and hence, when recently obtained, it is physically different from that which has been exposed to the contact of the atmosphere.

## Note VIII. p. 41.

Let $t$ denote the temperature of the anterior surface of the board, $\tau$ that of its posterior surface, $a$ its thickness, and $p$ its power of conducting heat. It is evident that, after the balance of supply and consumption has obtained, the quantities of heat continually dissipated at the posterior surface must exactly equal those accessions which are as regularly transmitted through the internal mass. But the discharge of heat from the surface is obviously proportional to its temperature, estimating this always by the excess above that of the room. That successive decrement of temperature will be expressed by $\tau$, which must therefore denote the momentary transfers of heat to the posterior surface of the board. To

express

express the equivalent internal communication, conceive the board to be divided into a number of parallel layers of a small, but determinate thickness. The measure of heat conveyed will depend on the joint consideration of the difference of temperature of two contiguous strata, and their conducting power. This successive decrement of temperature will be expressed by $\frac{t-\tau}{a}$, and therefore the momentary transfer is $= (\frac{t-\tau}{a})\,p$. Hence $(\frac{t-\tau}{a})\,p = \tau$, and $(t-\tau)\,p = a\tau$, and consequently $p = \frac{a\tau}{t-\tau}$. The formula is thus abundantly simple. When the thickness remains the same, the conducting power is directly as the temperature of the posterior surface, and inversely as the difference between the temperatures of the two surfaces. Instead of board, we may substitute a block of any solid materials; only the breadth must be large when compared with its depth, since no account is made of the heat which is spent at the edges.—But I need not stop to point out the application of those principles to practice.

## Note IX. p. 49.

Suppose the reflector LAM (fig. 5.) to be a small portion of a sphere, C its centre, and ACD its axis; and let the radiant object, situate directly in front at D, be a circle, whose diameter GH is equal to LM, the width of the reflector. This circle may be considered as equal to the concave surface of the reflector, since that surface is equal to a circle which has for its radius the chord AL, instead of DG or $\frac{1}{2}$LM; and, in small segments, the ratio of the chords and the corresponding sines approaches extremely near to equality. To determine the focal image, it is only necessary to trace, after their reflection, the concurrence of two rays that emanate from any point in the circumference of the radiating circle. The ray GCM,

GCM, which passes through the centre of the reflector, falling perpendicularly, will be sent back in an opposite direction; and the ray GA, which impinges at the vertex, will be reflected towards H, making an angle CAH equal to CAG. The point of intersection, K, is, therefore, the focus of G: and, in the same manner, it may be shown, that I is the focus corresponding to H. Hence the image thus formed, will likewise be a circle whose diameter is IK.—From this simple investigation, it follows that the radiant object and its image will subtend equal angles at the centre, and also at the vertex, of the reflector; for GCH = ICK, and GAH is the same as IAK. The light which falls upon the reflector is evidently concentrated in the focus after the proportion of $LM^2$ to $IK^2$, or that of $AD^2$ to $AF^2$. But it is an elementary proposition in optics, that the density of illumination is inversely as the square of the distance from the radiant point. Take OD = AF, and the density of the light received at O will be to that which is incident at A, as $AD^2$ to $AF^2$; wherefore, if the receptive object were transferred to O, it would be illuminated in the same degree by the direct afflux of light, as it was by reflection when it occupied the position at F. In other words, the intensity of illumination at the focus would continue unaltered, if the reflector were supposed to be converted into a simple radiating surface, of the same nature as the original circle GH.

I have, for the sake of simplicity, supposed the radiant to be a circle, and of equal dimensions with the reflector. It is obvious, however, that the above demonstration will apply to every other case; for the focal image will not have its density of illumination in the least affected by the change of magnitude of the radiating object, to which it is always similar and proportional.

## Note X. p. 73.

It will be sufficient for our purpose to calculate the measure of heat received at the centre of the screen, for this is the spot which acts principally upon the reflector, and the power of the surrounding heated space may be regarded as nearly in the same proportion. Nor will it alter materially the relation of effects, to suppose that the face of the canister is circular. Let C (fig. 9) be its centre, and CD a perpendicular meeting the screen in D; draw the radius CA, describe any circle BG and another *bg* indefinitely near it; and join AD, BD The heat sent from the point B, in the oblique direction BD, must have its intensity as the sine of the angle CBD; but it impinges with the same obliquity against the screen, and therefore, on both accounts, the impression which it makes at D will be as the square of the sine of CBD, or as $\frac{CD^2}{BD^2}$. And since the power of heat, under similar circumstances, must be always inversely as the square of the distance from its source; the true energy exerted at D is $= \frac{1}{BD^2} \times \frac{CD^2}{BD^2}$, or $\frac{CD^2}{BD^4}$. Put CD the distance of the screen $= a$, CA the semi-diameter of the canister $= b$, CB $= x$, and the ratio of the diameter to the circumference $= \pi$. Then the effect produced at D by the single point B being $\frac{a^2}{(a^2+x^2)^2}$, that of the circle BG must be $= \frac{2\pi a^2 x}{(a^2+x^2)^2}$, and consequently that of the infinitesimal ring $= \frac{2\pi a^2 x dx}{(a^2+x^2)^2}$. The integral of this expression is $\frac{\pi x^2}{a^2+x^2}$, which must therefore denote the influence of the circular space BG. Whence, because $\pi$ is constant, the whole effect of the face of the canister is proportional

tional to $\frac{b^2}{a^2 + b^2}$ or $\frac{CA^2}{AD^2}$, that is, to the square of the sine of the angle ADC.

When the screen is remote, the expression $\frac{CA^2}{AD^2}$ or $\frac{CA^2}{CD^2 + CA^2}$ may, without sensible error, be abridged into $\frac{CA^2}{CD^2}$. Hence, in this case, the calorific effect is directly as the surface of the canister, and inversely as the square of the distance of the screen.

## Note XI. p. 125.

Firmly persuaded of their general solidity, I gratefully adopt the leading principles of the very ingenious Abbé Boscovich. The capital work of that profound philosopher and elegant geometrician, entitled *Theoria Philosophiæ Naturalis*, a thin quarto, printed at Vienna about the year 1760, displays the happiest and most luminous extension of the Newtonian system. But it is not all of equal merit. The part which unfolds the fundamental views, and that which treats of mechanics and hydrostatics, are much superior to the rest. Chemistry, as a science, was yet in its infancy; and respecting the various intricate phænomena of corpuscular philosophy, the author, perhaps for want of better information, seems unfortunately to embrace only the earlier and cruder notions, which have long since been exploded. There are besides, either interspersed through the work, or appended to it, some obscure disquisitions, which might well be spared, since they contain only the sort of antiquated metaphysics that savours of the theologian. A neat abstract of Boscovich's Theory, would be a most valuable, and I presume, acceptable present to the public.

Fig. 10 represents the curve of primordial action. A, is

an elementary point, or particle, which exerts a certain varying energy on another point, supposed to be placed successively at different distances along the axis AB. The ordinates that stand above AB express attraction, and those which lie below it, denote repulsion. This axis and its perpendicular AD are, therefore, asymptotes to the extreme branches of the curve. The nearer portion must perpetually diverge from AB, to prevent the total collapse of matter, and oppose an insurmountable barrier to its penetration. The remoter branch of the curve, as it retires to a distance, will gradually assimilate itself to the law of universal attraction. The points, E, F, and G, of intersection with the axis, are points of quiescence: of these, E and G are stable, and F is instable. If, for example, a particle situate in E or G, be moved in the direction towards A, it will immediately feel repulsion and be forced again to recede; if it be drawn back, it must then experience attraction, which will solicit its return. The particle may thus oscillate about its centre, but soon must settle in the same position. On the contrary, if a particle in F be approximated to A, it will thenceforth become obedient to attraction; and if it be made to retire, it will be seized and transported by the power of repulsion. If once shifted, therefore, in the smallest degree from its place, it will, according to the direction which it has received, fly to E or G.

Fig. 10* exhibits the same curve, but with the modification which I have suggested in the text. It is a serrated line, whose gradations correspond to the breadth of the ultimate corpuscles, or the successive limits of action. Nature presents always individual objects, and proceeds by finite steps or differences. Absolutely continuous shades exist only in our modes of conception.

Note

## Note XII. p. 130.

The celerity with which vibrations are propagated through any medium, is proportioned to the square-root of its elasticity compared with its density. Professor Zimmerman, of Brunswick, found that salt water included within a very thick cylinder of iron, when urged by a force applied to the extremity of a long lever; and equivalent in effect to the weight of a column of similar fluid having a thousand feet in height, suffered a compression amounting to the 340th part of its bulk. But the atmospheric pressure is equal to that of an uniform column of about 28,000 feet high; or, if subjected to the additional weight of a column of one thousand feet, air would experience a contraction of the 28th part. Consequently the square root of $\frac{340}{28}$ or $12\frac{1}{7}$, which is very nearly $3\frac{1}{2}$, must express how much faster vibrations are sent through salt water than through atmospheric air.

This mode of experiment, though satisfactory in the gross, yet seems liable to error. In fact, the cylinder itself, notwithstanding its thickness, must likewise have suffered distension, which would thus augment the apparent effect. Mr. Canton's original experiments, where the water was equally compressed on all sides in the receiver of an air-pump, I consider as quite unexceptionable. The measure of contraction somewhat varies according to the temperature; but we may take it as a mean result, that the 25,000th part of the bulk corresponds to the pressure of a single atmosphere, or to that of a column of 34 feet of water. Therefore, since $\frac{25,000 \times 34}{28,000} = 30\frac{5}{14}$, the internal vibrations of water must shoot about $5\frac{1}{2}$ times swifter than those of air. The difference is thus even greater than was stated in the text. Hence, an impression would be transmitted through the ocean, from pole to pole, in the space of 175 minutes. Hence, too, is derived the intumescence of the sea, which commonly precedes a storm: for, suppose a hurricane to arise at the distance of 50 degrees, it will not reach us perhaps in less than 30 hours, while the agitation of the waters will begin to be felt in $48\frac{2}{3}$ minutes.

## Note XIII. p. 131.

Let ABDC (fig. 11) represent a beam of wood or bar of iron, laid horizontally with its extremities resting against two props. It will bend or swag by its own weight; and the curve which it thus forms, being gently and uniformly inflected, may be considered as an arc of a circle. The lower side is, therefore, extended, and the upper one equally contracted; but the particles of the middle stratum, though likewise affected in their general arrangement, retain the same mutual intervals and position. Hence, each layer will bear a strain proportioned to its distance from the centre of the beam's thickness, and the sum of all the longitudinal efforts must be as the square of the depth. With different degrees of curvature, those forces will be as the square of the quantity of depression; for the excess of an arc above its subtense, or the absolute strain, is, within moderate limits, proportional to the square of its sagitta.

Hence the reason why thin plates of wood or metal so easily bend, without suffering fracture. There are certain stones also which seem remarkably flexible: they are composed of thin layers, whose lateral adhesion is feeble, being divided by micaceous films. On the same principle depends the theory of cordage; for the parallel fibres should act separately, and excessive twisting makes a rope stiff and apt to break.

In the case of a solid beam, the upper parts, being condensed, exert repulsion; and the lower, being distended, acquire attraction; and these opposite forces tend both equally

to

to restore the original figure, by producing an effort which counterbalances the action of their own weight. To determine the precise effects, let G, H, and I, (fig. 12) denote three adjacent particles of the lower stratum: the forces HG and HI by which the particle H is attracted, are resolved into HK, KG, and HK, KI; of which KG and KI destroy each other. Hence the strain is to the reaction occasioned by a single particle, as GH to HK, or as MH to GH. But this reaction being only an equipoise to the pressure of the stratum, the longitudinal strain must be equal to the number of particles that would be contained in HM, or to the weight of a similar stratum, having for its length the diameter of curvature. But it was observed that the internal parts, according as they approximate to the middle line, are proportionably less affected. Consequently the mean strain of the whole beam is measured only by the radius of curvature.

Put $l =$ length of the beam, $h =$ its depth, and $a =$ its quantity of depression. Then $\frac{l^2}{8a}$ will denote the radius of curvature, and $\frac{2ah}{l}$ will express the contraction and equal distension which are produced. Hence the rate of compression is $= \frac{l^4}{16a^2h}$ or $\frac{1}{h}\left(\frac{l^2}{4a}\right)^2$. Thus, in the case of deal, $l = 138$ inches, $h = .45$, and $a = 2.5$; therefore $\frac{1}{.45}\left(\frac{138^2}{10}\right)^2 = 8{,}059{,}500$ inches, or 671,625 feet. Consequently, under the pressure of a column of similar materials and a thousand feet high, a fir board would suffer a contraction equal to about the 672nd part of its length.

## Note XIV. p. 131.

Not to multiply quotations, I shall select only the most striking.—

Σῆμα δέ τοι ἀνέμοιο καὶ οἰδαίνουσα θαλασσα
Γιγνέσθω· καὶ μακρὸν επ' αἰγιαλοὶ βοόωντες,
Ακται τ' εινάλιοι, ὁπότ' εὔδιοι ἠχήεσσαι
Γίγνονται, κορυφαί τε βοώμεναι οὔρεος ἄκραι.

ARATI PHÆNOM. 177-180.

Continuo, ventis surgentibus, aut freta ponti
Incipiunt agitata tumescere, et aridus altis
Montibus audiri fragor; aut resonantia longe
Litora misceri, et nemorum increbescere murmur.

GEORG. I. 355-358.

——— Ceu flamina prima
Cum deprensa fremunt silvis, et cæca volutant
Murmura, venturos nautis prodeuntia ventos.

ÆN. X. 97-99

The same idea is perhaps more nobly painted by Thomson.—

Ocean, unequal press'd, with broken tide
And blind commotion heaves; while from the shore,
Ate into caverns by the restless wave,
And forest-rustling mountains, comes a voice
That, solemn sounding, bids the world prepare.
Then issues forth the storm with sudden burst,
And hurls the whole precipitated air
Down in a torrent.—

WINTER, 148-155.

## Note XV. p. 133.

The influence which the slow communication of impulse through the atmosphere must have in heightening the effects of any casual disturbing force, receives illustration from the phænomena of tides in narrow seas, where the waters are observed to rise far above the height assigned by theory. Compare, for example, the prodigious accumulation which takes place in the British Channel, with the moderate reciprocating swell that prevails in the free expanse of the Pacific Ocean. Straits and estuaries, by confining the current of influx, cause a derangement similar to what is produced by the imperfect sympathy between the distant portions of the air, in augmenting the unequal distribution of that fluid. If our globe had been smaller, the variations of the barometer would have been proportionally diminished.

To investigate accurately, therefore, the origin and effects of wind, it is requisite to consider the motive forces not as acting simultaneously, but as spreading themselves with a progressive diffusion. The problem will hence depend, for its complete solution, upon the extension of the method of *partial differences*; a discovery in the higher calculus to which it first gave rise. And though, in an aqueous medium, the actual motion is much slower and the propagation of impulse swifter, the currents of the ocean must likewise experience a certain degree of modification. The profound researches of Laplace on that subject would consequently require some revision.

## Note XVI. p. 136.

Mr. Hume is the first, as far as I know, who has treated of causation in a truly philosophic manner. His *Essay on Necessary Connexion* seems a model of clear and accurate reasoning. But

But it was only wanted to dispel the cloud of mystery which had so long darkened that important subject. The unsophisticated sentiments of mankind are in perfect unison with the deductions of logic, and imply nothing more at bottom, in the relation of cause and effect, than a *constant and invariable sequence*. This will distinctly appear from a critical examination of language, that great and durable monument of human thought. Etymology has indeed been often exposed to ridicule, by the crude and fanciful opinions of philologists and dreaming antiquaries. Yet therefore to cover it with unqualified contempt, would only betray ignorance. To trace etymologies with sober circumspection, and guided by the light of philosophy, is not only a liberal exercise of ingenuity, but elucidates finely the various phases of the human mind, and represents to our view the history and progress of its more abstruse operations. Derivations are not safely inferred from solitary instances; they must be drawn from the comparison of whole classes of words, and the uniform analogy of different languages. It would be foreign to my present object to engage in such discussions. I trust, however, that the few examples which I shall select will amply confirm what has been advanced.

*Ursach*, in German, is the appropriate term for *cause*. The same word, with only slight alterations, runs through the several branches of the Gothic stem. It is compounded of *ur*, an inseparable preposition, and *sache*, a substantive noun. *Sache* denotes *a thing of moment, an interesting and important object*. The prefix, *ur*, signifies *before* or *anterior*. It now occurs only in composition, but its radical force is there clearly marked. By the German mineralogists, it is employed to designate the supposed primitive substances: Thus, *Ur-trap* and *Ur-kalkstein*—comprehended under the general class of *Uranfangliche Gebirgsarten*. The same particle had passed into other dialects, and is even retained in English, though now very seldom used except by the poets—" *Ere* the mountains were

were formed"— *Erst* is evidently the same preposition in its superlative degree; in German, it means *first* in the order of succession; in English, it had a kindred signification, but has become obsolete.—Hence, combining its elements, the term *ursach* expresses merely *the capital object which precedes.*

It is curious to remark the shades and transitions of the word *sache.* It comes to signify *an affair, a subject of dispute, a pleading or law-suit.* The plural, *sachen,* denotes *goods or effects.*—In Swedish, *sak* has a corresponding extension, and its plural, *saker,* likewise signifies *moveable effects.* The English noun *sake* is of the same origin, but expresses more generally *whatever concerns us.* The verb *forsake* reflects a similar idea.

The Greek Αἰτία and the Latin *Causa,* correspond exactly to the German *Sache.* They had come to denote that more limited object—a law-suit. But there are some traces of their primitive sense. Thus, *cosa* in Italian, the same with *chose* in French, means *thing in general.* We observe also a similar progress; for *causer,* or in old French *choser,* means to *talk,*—the gradation of *causas dicere.* (The verb *causare* had in Latin corresponded exactly to *causer*; as appears from this line in the ninth eclogue—*Causando nostros in longum ducis amores.*—) A like transition has taken place in the northern languages. *Ding,* in German, assuming the aspiration peculiar to the Anglo-Saxons, passes into the English *thing.* In Swedish, the same word is *ting,* which besides its original signification, denotes a *trial, or a seat of justice.*—*Causa,* therefore, means simply *an object of importance:* the idea of priority or concomitance is but implied The correlative term, *effectum,* which is of later origin, marks the sequence. In the plural, it corresponds, in Latin, French, and English, to *sachen.*

The other words, used as synonimous with *cause,* mark a similar *antecedence. Ground, principle, origin*—all express the order of the succession of events. In Swedish, the appropriate term is *uphof,* which literally means to *heave up,* in allusion probably to the process of germination.—*End, purpose, design, intention,*—these refer to the sequel, to the result of the successive

cessive concomitancies.—To *produce*, is to *bring forward*, and denotes the continuity of events.—*Exposition*, *explication*, *explanation*—merely express our mode of conceiving the sequence or concatenation. To *account* for an appearance, is to enumerate the several links of the chain.

Between conjunctions and prepositions, the distinction is arbitrary and accidental. When they signify precedence, conjunction, or proximity, they are fit to express causation. From *pro*, which means *before*, come *prope* and *propter*.

"*Propter* aquæ rivum, sub ramis arboris altæ."

*Propter*, therefore, denoted proximate anteriority, but came afterwards to be more generally used as a causative conjunction. *Propterea*, is *nigh before those*.—*For*, *therefore*, *wherefore*, express the same idea. Hence also the *pero*, *perche*, or *porque*, of the Italians, and the *pourquoi* of the French. The Latin *quamobrem* is equally a compound.—*Luego*, in Spanish, the same as *loco (ipsissimo loco)* denotes immediate succession. The Spanish *pues*, and the French *puis* or *puisque*, derived from *post* or *postea*, mark a distinct sequence. *Quare (with which thing)* signifies mere contiguity, and hence *cur*, and, in French, *car*.—*Why* and *where* are of the same descent, being formed from the German *wo*, which, in the Anglo-Saxon, was *war*: *warum*, the corresponding word in German, is by its composition exactly *where-about*.—*Seit*, which means *by the side of*, and was equivalent to *since*, has separately become obsolete in German; and so has also its representative in English, *sith*. But, with the addition of the pronoun *dem*, or *this*, it is actually used as causative, and is synonimous with the French *puisque*. It is *sedan* in Swedish, from which dialect it seems to have passed into the English,—*sithance*, afterwards shortened into *since*.—The Swedish *med* is the same as the German *mit*, signifying *with*: hence *emedan* is equivalent to *since*, or the French *parceque*.—The Latin *cum*, denoting merely coincidence of time, is often rendered by *since* or *because*. *Then* has in English the same double application; and *thence*, probably *then-this*, seems only a pleonasm.—About the age of Charlemagne,

Charlemagne, *this* and *thas*, in German, were equivalent to *because*. *Sunder* (asunder) and *nouen* then corresponded to *but*, as the latter word, or *now*, does nearly so at present.—I might easily pursue these illustrations, but it would be superfluous. Enough has been said to prove that, in every language, the casual conjunctions were, or still are, only prepositions expressive of contiguity or succession.

But in conceiving the relation that subsists between cause and effect, do we not *feel* something more than the mere invariable succession of events? I will admit the fact, but I maintain that, like many other spontaneous impressions, it is a fallacious sentiment, which experience and reflection gradually correct, yet never entirely eradicate. It is a vestige of that extended sympathy which connects us with the material world: It is the shade of that propensity of our nature to bestow life and action on all the objects around us; to clothe them with our own passions and habits, and to discover the image of ourselves reflected from every side. This disposition is very conspicuous in children; nor is it even wholly effaced by the progress of age. Hence the true foundation of what is called figurative language. Vivid imagery always implies a real, though transient, belief. Personification is the most familiar either to those not accustomed to repress the spontaneous emotions, or to those who have cultivated the power of recalling the passions in all their native glow. A choleric man, who happens to strike his foot against a stone, vents his rage on that obstacle, because, for the moment at least, he actually believes it to be animated like himself. The efforts of the poet and those of the philosopher, are diametrically opposite. The one endeavours to subdue the passions, and to correct our early and false impressions; the other seeks to renew our infant visions, and to expand the warm and illusive creation of untamed fancy. Yet, after a severe exercise of reason, the mind finds grateful relief in that magical and fantastic colouring which

which tricks external objects, and diffuses life and sentiment throughout nature. Pomp of language—smoothness and harmony of verse—are only the accessory decorations; fervid animation constitutes the soul of descriptive poetry. It is hence that mythology, the religion of the vulgar, has ever been a favourite subject with the poets.

These observations are amply confirmed by the structure of language. Our senses are first aroused by the changes that take place among the surrounding objects; and upon the succession of those changes, is all our experience founded. Object—motion—object,—such is the series in the perpetual concatenation of events. Hence, corresponding, are the primary and essential elements of speech—the verb, and the noun. In early periods of society, every object was viewed as animate, and consequently distinguished by sex. Originally, therefore, every substantive noun was referred either to the masculine or the feminine gender. As the passions cooled and knowledge advanced, the spontaneous belief of animation became blunted or effaced, and the distinction of gender in language gradually fell into disuse. The neuter is evidently of a later growth. In Latin, the words that come under this description are generally derivative—have often been altered and improved, which they betray by their mutilated terminations—and sometimes have been introduced from foreign countries, in the progress of refinement. Gender has been entirely banished from our own language; which, for that reason, is perhaps the most philosophical in Europe. Yet the poets and the less instructed classes of men, are still accustomed, in speaking of certain inanimate objects, to bestow sexual distinction.

It is the characteristic of the verb, to denote *motion*. All verbs in fact appear, when analysed, to involve the idea of action or transition. To this principle there is no real exception; for even such as are now employed to signify modes of rest or position, have originally expressed only the peculiar motions

motions which preceded those states of existence. I need not go far in search of illustrations. *To be* and *to do* are evidently of synonimous import. The equivalent word in Latin is *ago*; in Greek, *εἶμι*, *I run*; and in French *porter*, to *carry*;—all of them expressing actions which mark existence.—The verbs styled neuter were only distinguished at a later period of society. Thus, I conceive that *to sit* meant originally *the act of seating oneself*, which is now expressed by *to set*. In the same manner, *to lie* was at first equivalent to the verb *to lay*; and accordingly these terms are often confounded by the vulgar.—Those verbs employed to signify the various affections of the heart, perhaps only denoted the gestures and movements which represent their corresponding sentiments. The more abstruse mental operations are expressed by the help of external similitudes.—In the progress of language, verbs are sometimes formed from nouns, and nouns derived from verbs. Motion may be characterized by its object, or the object reciprocally may determine the nature of the motion with which it is associated. *Thing* and *factum* are derived from the verbs *thun* and *facere*; πρᾶγμα, χρῆμα, ἔργον are equally derivative; and *res* was perhaps formed from *reor*, which primitively signified *to act*. In like manner, the verb *to think* seems to be of the same descent as the noun *thing*.

In farther illustration of these views, I would observe, that, in all the cultivated languages, the words, *power*, *force* and *energy*, had denoted originally mere corporeal exertion. *Force* is evidently derived from *fortis*, force and bravery being esteemed synonimous. And *vis*, as it forms the genitive *viris*, was probably akin to *vir*, like *viritas*, and expressed *manhood*. *Virtus*, which is of the same descent, was appropriated to *bravery*, as the quality most distinguished in rude ages.

From motion, seem derived our ideas of time and space, which are often interchangeable terms. The German word *zeit*, denoting *time*, was at first expressive only of *motion*: for, in Swedish, it has passed into *tid*, the same with the English

*tide*.

*tide*. The primitive sense of *tide* may be gathered from its compounds, *noon-tide, betide, tidings*, &c. " Roll back the tide of time," is but a repetition of the very same idea. The present application of *tide* to denote the flowing of the sea, in like manner as *fluxus* from *fluo*, bespeaks a common origin. That word signified flowing in general, and was thence transferred to time and tides; its former application has been retained in Germany and Sweden, its latter in England.—*Tempus* refers merely to the fleeting aspect of the sky; and hence *tempestas, tempestiva*, and its modern derivatives.—χρόνος has perhaps originated from ῥέω, *curro*.

*Matter*, a term at present so abstract, signified in its primitive import merely *timber for building*. Thus ὕλη in Greek and *sylva* in Latin, denote literally a *wood* or *forest*. *Materies* is exactly equivalent to the English plural *materials*. The same word in Portugese is *madeira*, whence the name of the island discovered by those early and adventurous navigators. It is the property of composition to which the mind refers.

## Note XVII. p. 152.

This *luminous air* is what experimenters usually term the *electric spark*, as I have clearly shown in a paper written so long ago as the year 1791, and which subverts some favourite theories in electricity. That single point is pregnant with curious consequences. Though fully assured of the justness of my conclusions, I have more than obeyed the advice of the Roman critic—*nonum prematur in annum*. But I design soon to revise it, and extend my experiments and reasonings into that beautiful branch of science.

Note

## Note XVIII. p. 166.

The chemical philosophers were the first who considered Heat as a diffusive subtile fluid, subject to partial derangement, yet constantly endeavouring to maintain its equilibrium. Boerhaave taught, that it was distributed among bodies according to their respective quantities of matter; and this opinion, being countenanced by the system of universal gravitation then coming into vogue, seems to have been long currently received. But after the art of experimenting was more cultivated, the inaccuracy of the hypothesis became quite apparent. In some cases, the aberration from the law of density, is most striking. Thus, mercury contains not the thirtieth part of the quantity of heat which is lodged in an equal weight of water. Heat is, therefore, distributed among adjacent substances, by proportions peculiar to each recipient. This quality has been termed their *capacity*; and the circulation of heat is compared with that of a liquid poured into a system of connected vessels having different diameters. It will quickly rise to the same level in them all, and consequently each addition must become shared among them in proportion to their respective sections. The thermometer is a branch of that general system, and marks the height of the supposed communicating liquid.

Such is the ingenious theory founded by Irvine and Wilcke, and afterwards improved by Crawford, Gaodilin, and others. It is perspicuous and comprehensive; and it accords perfectly with the phænomena. The relative absorbent disposition of various substances being once ascertained, the several circumstances regarding the distribution of heat are thence reduced to a very simple calculation. Yet, without questioning the correctness of its results, I would observe, that, in the framing of this theory, too much has been sacrificed to popular illustration. Capacity was a term unfortunately

 chosen,

chosen, because it seems to convey an idea, that heat received into a body serves merely to fill up the pores or internal vacuities. But heat is evidently not passive; it is an expansive fluid which dilates in consequence of the repulsion subsisting among its own particles; and it would spread indefinitely through space, if it were not fixed or retained by the counterbalancing attractive power of the substances which absorb it. Were each corpuscle to exert the same action, this universal fluid would be disseminated among bodies exactly in proportion to their respective quantities of matter. The mutual adhesion depends, however, on the density of the substance, modified by its degree of inherent disposition to combine. A sort of affinity is thus produced, corresponding in effect to *capacity*, and which I have denominated by the phrase "*specific attraction for heat.*" This attraction is manifestly determined by the peculiar nature of each body. To trace its immediate origin, is not more possible perhaps than to discover the source of other physical properties. Yet there appears some tendency towards a general principle: the particles of heat, like those of all expansive fluids, have their repulsion diminished in proportion to their mutual distance; while the molecules of the containing substance suffer the corresponding decrease of attraction after a slower ratio than the spaces of internal separation. Heat has, therefore, a narrower range of density than the bodies with which it combines. It holds a sort of middle station, and is distributed according to the quantity of matter, joined to the consideration of the space which this occupies; that is, it obeys some compounded relation of the weight and the bulk. Hence the denser bodies receive a proportionally smaller share of heat. Thus, a pound of metal contains less heat than one of stone; this, less than an equal weight of liquid; and this last, still less than a pound ofany species of gas.

When two bodies are united chemically, the compound has an attractive force generally different from that of the mean result.

result. Hence a corresponding portion of heat is, during the act of coalescence, either absorbed or evolved. Thus, water, on being joined to sulphuric acid, occasions an extrication of heat, because the diluted acid exerts less power of adhesion than did its ingredients. And, for an opposite reason, the muriate of ammonia, in dissolving, is attended with an absorption of heat, or an apparent production of cold. But for the most part in composition, there is a loss of attractive energy, and consequently a disengagement of heat. On the same principle, depends the extrication of heat which takes place during the process of combustion. The diminution of attraction is here so violent and sudden, that part of the heat discharged is finally projected in the form of heat

Nor are those changes peculiar to the chemical combinations only. Every substance capable of assuming different states of constitution, betrays likewise analogous variations of attractive force. When a solid body melts into a fluid and thence passes into vapour, each transit is marked by an augmentation of that force, and is therefore accompanied with a corresponding absorption of heat; during which process, the temperature must evidently remain stationary. Thus, a lump of ice, transported intensely cold into a close apartment, will grow warmer by regular gradations, till it begins to thaw, and there the farther accumulation of heat will appear to be suspended. And if the water so formed be poured into a covered pot and set over a steady fire, the temperature will again rise uniformly, till it reaches the limit of boiling, when the act of conversion to steam will henceforth absorb the whole affluent heat; yet the temperature will mount still higher, if the escape of the vapour be prevented, but which soon acquires such prodigious elasticity as to burst whatever obstacle can be opposed to it.

These curious facts have been long observed. They were certainly known soon after the middle of the seventeenth century. On the stability of the points of freezing and boiling,

 indeed,

indeed, depends the mode of graduation adopted by Reaumur for his thermometer, and afterwards applied more successfully to the mercurial thermometer of Farenheit. But it was reserved for the celebrated Dr. Black to examine the phænomena of the passage to fluidity with accurate attention; and had that able chemist discerned all its consequences, or possessed the boldness and ardour of his rivals, he would have deserved our unqualified praise. But he satisfied himself with taking a partial glimpse of the subject, while the theory of *capacity*, at once so luminous and comprehensive, was reared without his participation. Not daring to reject the system of Boerhaave, he sought only to correct it. That correction was however insufficient; for, besides the quantity of heat absorbed in the passage of ice to water, the fluid henceforth requires larger additions, than it did in its former state, to produce the same ascents of temperature. If the term *latent* expressed merely a simple fact, it might be admitted; but, viewed in opposition to its correlative, *sensible*, it conveys an erroneons notion, as if heat lodged in a body consisted of two distinct species or modifications. A thermometer measures only the heat contained in its own bulb; any farther indication is inferred by an act of reasoning. The ideas associated with the expression *latent heat* have spread a cloud of mystery and paradox most unfavourable to the progress of real science. Notwithstanding the obstinacy of habit, they are indeed falling into discredit, and must soon melt away into a system which is far more general, consistent, and philosophical.

## Note XIX. p. 170.

This fine discovery was made by the ingenious Dr. Irvine. The solution of the problem is derived from the measure of heat absorbed or evolved in certain changes of constitution, compared with the corresponding alterations of specific attraction. Two distinct methods have been employed; 1. from the heat which disappears on the affusion of boiling water

water upon snow,—and, 2. from that which is extricated when water is mixed with sulphuric acid at the same temperature. Their *capacities* before and after the transition or combination are ascertained by observing the different effects of the admixture of siliceous sand or small lead-shot. To determine that of snow, is on several accounts the most difficult; and I should therefore prefer the second mode of investigation. Nor is even this altogether exempt from uncertainty, for the specific attraction of a substance, while yet under the same form, is perhaps not strictly constant through its whole extent of temperature. The passage to fluidity or to vaporization may be prepared by successive though insensible gradations.

## Note XX. p. 174.

I shall here mention a ready mode for discovering the relative affections of the several permanent gases with respect to heat.—Their *capacity* is increased by rarefaction, and hence a corresponding portion of heat becomes again evolved when they recover their former state. Having therefore fixed a delicate thermometer in the centre of a large receiver, extract most of the air, leaving perhaps only the tenth or hundredth part, and allow the apparatus to acquire exactly the temperature of the room. Then suddenly admit the air into the partial void, and the heat now disengaged will proportionally raise the general temperature. Repeat the exhaustion, but after the necessary interval of time, open a communication with some other species of gas: the same quantity of heat will be liberated as before, but its effect may be different. If the gas be more absorbent of heat than an equal bulk of common air, it will experience less alteration of temperature. Hence their order of arrangement is ascertained; though to determine the true relation, would require some farther research. The heat thus suddenly set loose is not all exerted upon the contained gas; the greater part of it is spent in warming the internal surface of the receiver. This expenditure, however, being obviously proportioned to the relative extent of surface, might be discovered by repeating the observation with another receiver of a similar form but much smaller dimensions. Hence, by a simple computation, the *capacity* of the gas will be derived.

In the case of hydrogenous gas, no calculation was required; for, on its admission, it suffered exactly the same change of temperature as atmospheric air. Hence, in the same space, they both contain equal measures of heat; which agrees very nearly with Dr. Crawford's experiments.

## Note XXI. p. 176.

It is well known that fluids are projected from small orifices with a celerity proportioned to the square root of the height of the incumbent column. Under the same pressure, therefore, the celerity corresponds to the inverse subduplicate ratio of the density. Hence we derive an elegant method for ascertaining the specific gravities of different gases. Let a tall cylindrical receiver be fitted with a brass cap and stop-cock, and terminated with a short pipe having an orifice of about the 50th part of an inch in diameter: with the point of a diamond, draw a graduated perpendicular line along the side, and set it upon a very wide basin full of water, which, by applying the mouth to the pipe, may be sucked up to some moderate height, and there kept suspended by again turning the cock. Fill a large bladder with any sort of gas and tie it to the pipe; now suddenly open the cock, and observe carefully the number of seconds which the water takes to descend to a certain intermediate mark.—Thus, the whole height of the column being 10 inches, I found that when common air was administered, it took 130 to fall through the space of one inch;

inch; but, on supplying hydrogenous gas, it made the same descent in 45″. And $(45)^2 : (130)^2 = 9^2 : (26)^2 = 1 : 8\frac{1}{3}$ which denotes the comparative rarity of the gas.

## Note XXII. p. 177.

The velocity with which air would rush into a vacuum, is the same as that of projection under the equiponderant column. But the mean pressure of the atmosphere is equal to the weight of a column of uniform density and about 28,000 feet high. That velocity is therefore $= 8\sqrt{28,000} = 1329$, or in round numbers 1350 feet per second nearly.

## Note XXIII. p. 177.

Water is indeed about 900 times denser than air at the mean temperature; but then it has a smaller *capacity*, or is reckoned to contain heat only in the proportion of 5 to 9.

## Note XXIV. p. 181.

The mean density of the earth deduced from the observations made by Dr. Maskelyne on the sides of Schehallien, an insulated mountain in Perthshire, is $4\frac{1}{2}$, reckoning water as usual the standard of comparison. Mr. Cavendish has lately assigned a greater quantity, or about $5\frac{1}{2}$, from a very elegant experiment on the principle of torsion, which Coulomb employed so successfully in a variety of delicate researches. Perhaps the true proportion would be found to lie between these limits. The observations of Dr. Maskelyne, however skilfully conducted by that eminent astronomer, were performed under the most unfavourable circumstances, in a foggy climate and a rainy season. And Mr. Cavendish's experiment was not perhaps made on as cale sufficient to afford very great precision. Nor is it at all improbable that the apparent force of attraction was in some degree augmented by a slight infusion of magnetic virtue; for the masses of lead which discovered their mutual appetency, might yet contain a certain admixture of iron in a state of such intimate combination as to resist the action of chemical solvents.

I am disposed to think, that, instead of selecting a conical hill with a view to ascertain the deviation of the plummet, it would be more eligible to place the observer successively on the opposite sides of a narrow vale, bounded by two ranges of lofty mountains which run from east to west. Those stations would be very commodious for determining the altitude of a star, and their true distance could be found trigonometrically with the utmost exactness. The mountains themselves might be surveyed by considering them as composed of a number of parallel and vertical slices formed by planes in the direction of the meridian. The best scene, that I am acquainted with, for attempting these operations, is in Upper Valais, where the Rhone holds a westerly course, and the enormous Alps, in a double chain having more than a mile of perpendicular height, approach at their lower flanks perhaps within two or three miles.

## Note XXV. p. 181.

This opinion that the earth is growing continually warmer, stands directly opposed to the favourite hypothesis of the celebrated Buffon. That eloquent and fascinating author, who possessed such unrivalled powers of description, was misled by the illusions of fancy and the vague ideas which then prevailed concerning the properties of heat. To attempt any serious refutation, however, would be quite superfluous.

But the temperature of our globe increases with a progress extremely slow. No mighty change has actually taken place within the period of authentic history. Yet, from the

the concurring testimony of ancient writers, it seems unquestionable, that the climate, over the whole of Europe, has assumed a milder character. The severity of winter, as felt in the age of Augustus on the borders of the Danube, can only be compared, in our own times, to what obtains on the banks of the Vistula or the Neva. The only apparent exception, perhaps, to this remark, is the present degraded state of Iceland and Greenland, which were discovered and planted by the roving pirates of the North. These settlements are represented to have been once comfortable and even flourishing: but the colony of Greenland has long become extinct, and the scanty population of Iceland is now sunk into the lowest state of wretchedness, unable to struggle with inclement skies and a penurious soil.—We should, however, make great allowance for the heated imagination and very limited experience of those bold adventurers who issued from the rocks and dark forests of Scandinavia. As long as fresh recruits continued to arrive, they fanned the ardour of active enterprize; but after the tide of emigration was spent, the original colony began to languish and did gradually decline; and in the lapse of a thousand years, the tardy melioration of climate has been unable to keep pace with the devouring influence of political decay.

But the history of our species, with all its busy and tragic scenes, shrinks into a point in comparison of those vast cycles which are familiar to the mind of the geologist or astronomer. Man, the last and most perfect of Nature's works, is only a recent inhabitant of this globe. The survey of its structure transports us far beyond the origin of animated beings, and even the incipient germs of vegetation, into the fathomless depths of primæval Time.—I am aware that cosmological speculations, being so wonderfully seductive, and not capable of demonstration, may lead to extravagance and absurdity. Yet there are certain characters imprinted on the surface of our planet, which it seems impossible to mistake. Of this kind

kind I consider the marks of progressive warmth and amelioration of climate. It is well known that, in the same parallel of latitude, the degree of cold which prevails at any place depends on its altitude above the level of the sea. At some assignable height, therefore, must commence the reign of perpetual congelation. But this limit, in all probability, descended at a former period much lower than it does at present. To prove this fact, I shall mention certain appearances which forcibly struck me in the course of a most agreeable tour, performed on foot, in the year 1796, through the mountains of Switzerland, in company with my invaluable friend Mr. T. Wedgwood, whose discerning eye and cooler judgment might well inspire confidence.—

In crossing Mount Grimsel, from Obergestlen, near the source of the Rhone, to Guttanen, and thence into the charming vale of Hasli, we remarked that the rocky surface, over which the road lay, had changed its features before we reached the middle of our descent. The vast bed of granite which covers the summit and flanks of that mountain, affects, at the upper part, to swell into a succession of broad convexities, so very smooth that it has been found expedient to cut small transverse gutters, for the sake of giving a firm footing to the mules and beasts of burthen. This appearance continues to a considerable distance below the Spital, which stands on the verge of a small gloomy lake, the source of the Aar. But the rock now begins to present a rough surface, being marked with numerous waved furrows, evidently worn in the lapse of ages by the action of descending rills. In the higher region of the mountain, a thick covering of snow would absorb the accidental rains, and even while it softened and partly melted away, it would gently and uniformly spread the icy water over the bottom, and therefore defend the granitic crust from erosion. But the Grimsel is never at present covered the whole year with snow. The line of perpetual congelation runs several

veral hundred feet above the highest part of the passage, and consequently, before the tortuous furrows were scooped out, it must have descended at least 2000 feet lower than its present station. At that remote period, the lake which now feeds the Aar was probably a glacier, like that of Furca, the common parent of the Rhine and the Rhone.

The formation of glaciers has not yet been explained in a satisfactory manner. It is even disputed whether those seated amidst the Alps are on the whole in a state of increase or of diminution. I shall, therefore, throw out some hints which may help towards a theory of this curious and interesting subject.—The line of congelation in any latitude is not absolutely fixed, but fluctuates with the progress of the sun, between certain limits. In winter, it descends below the mean position, and rises, in summer, as much above that. Under the tropics, where the heat is almost uniform through the whole year, those extreme boundaries must approach very near to each other. But in the higher latitudes which experience a very considerable diversity of temperature during the revolution of the year, the space subject to the alternate process of thawing and freezing must have its breadth proportionally great. The snow which falls in winter being partly dissolved by the summer's heat, becomes soaked with humidity, and congeals into a solid cake on the return of the severe season. Thus a zone of ice gradually collects. This production, in the bosom of the higher Alps, forms extensive plains, termed *mers de glace*; but, on the sides of those towering mountains, it gives rise to *glaciers*. The snow which falls above the superior limit of congelation, from its powdery and incohesive quality is incapable of much accumulation: loosened by the impression of the sun, it slides down, and gathering force in its descent, it often precipitates itself in those dangerous *avalanches*. But I consider glaciers themselves as formed only by *avalanches* of a rarer and more formidable kind.

The

The icy zone will accumulate, till its weight at last overcomes its cohesion; then giving way, it will rush down the side of the mountain with irresistible sweep, and spread its shivered fragments. This statement agrees with the phænomena, and explains the reason why glaciers are not observed among the Andes.

But those vast irregular masses of ice, thus transported from their native seat, and planted in a lower and therefore warmer region, must now suffer a continual diminution. This progress, however, seems to be extremely slow. Several hundred years may be required to melt the whole away; and in the mean time, a new magazine of ice is again collecting between the boundaries of perpetual congelation. Hence the production of glaciers and their subsequent decay, are events necessarily connected together, and must succeed each other at regular periods.

The gradual contraction of glaciers is rendered visible by a circumstance worth mentioning. Small fragments of rock are detached from the sides of the schistose mountains, and being successively transferred along the sloping surface of the ice, they are finally deposited at its lower margin. Hence the origin of *maremes*, or those mounds of loose rough stones which are seen heaped up at some distance from the edge of a glacier. The progress of such accumulation and the gradual retreat of the icy boundary, are distinctly marked; for the range of hillocks appears sprinkled with natural pines, of a size or age proportioned to their remoteness, the nearest only just emerging from their seed. This remarkable fact is known to every tourist who visits the glacier of Grindlewald.

But I regard it as highly probable, that the hillocks so frequent in Upper Valais are of the same nature. They seem to rise like islets in the plain; they are generally clothed with wood, and often show vestiges of ancient turrets. They consist of loose materials that have evidently been detached from the

the neighbouring rocks, their angles sharp, and their surface rough, without any marks whatever of attrition. There was perhaps a time, therefore, when glaciers descended more than 3000 feet below their actual line, into the bottom of the valley. Above Sion, on the road to Sitter, the string of wooded hillocks first attracts the eye of the traveller. The mean heat is only about 9 degrees; yet in summer the atmosphere feels sultry and oppressive. Every object around begins already to assume the character of an Italian climate:

Sole sub ardenti resonant arbusta cicadis.

## Note XXVI. p. 186.

Let $a$ denote the emergent angle, and that of inflection, and $m$ the attractive force which is exerted by the luminous body. The effect is similar to what takes place in refraction, and consequently $m$ sin. $a$ = sin. $\alpha$. Taking the differentials, $m\,da$ cos. $a = d\alpha$ cos. $\alpha$, and hence $\frac{d\alpha}{m\,da} = \frac{\text{Cos. } a}{\text{Cos. } \alpha}$. But $da$ is constant, because the emergent rays are supposed to spread uniformly; therefore $d\alpha$, which must express the breadth or density of the inflected pencil, is proportional to $\frac{\text{Cos. } a}{\text{Cos. } \alpha}$.

## Note XXVII. p. 192.

The equation $v\,dv = f\,ds$ expresses the general relation which connects the velocity, the space, and the actuating force. But, in every species of vibratory motion, this accelerating or retarding force is proportioned to the distance from the point of final quiescence. Therefore $v\,dv = -s\,ds$, and consequently $v^2 = -s^2$; that is, the increase or decrease of the square of the velocity is as the difference between the squares of the respective distances from the centre. Hence if the scale of distance be enlarged, the corresponding celerities

rities will be likewise augmented in the same proportion, and of course the time consumed in those similar motions must remain unaltered.

Let $S$ denote the limit of evagation, then $v = \sqrt{S^2 - s^2}$, and $dt = \frac{ds}{\sqrt{S^2 - s^2}}$; whence the integral is, $t =$ arc sin. $\frac{s}{S}$. Therefore, if a semicircle be described on the line of oscillation, and perpendiculars drawn from any points of it, the number of degrees contained in the arc thus intercepted will represent the time elapsed.

## Note XXVIII. p. 197.

In gaseous fluids, the vibratory and the projectile energies are very nearly related, being indeed only modifications of the same force. The celerity displayed, whether actual or potential, depends on the comparative elasticity of the medium, and which is measured by the altitude of the equiponderant column. Let $n$ denote the number of particles in that column, $a$ their mutual distance, and $g$ the power of gravitation. Then $gn$ will express the incumbent pressure, which, being exerted through the space $a$, generates the velocity of projection. But, by the general theorem in dynamics, $v\,dv = f\,ds$, and therefore, in the present case $\frac{1}{2}v^2 = gn \times a$. If a particle fell through the whole height of the column, or $na$, by the sole action of gravity, we should have $\frac{1}{2}v^2 = g \times na$, the same as before. And thus the velocity with which a gaseous fluid rushes into a vacuum is equal to that acquired by a perpendicular descent through the equiponderant column.—To discover the celerity of vibratory transmission, put $\alpha =$ the minute displacement which each particle suffers, or the breadth of the tremor excited. The elastic force developed is, therefore, $= \frac{\alpha}{a} \times gn$, and hence $v\,dv = gn \times \frac{\alpha}{a} \times d\alpha$,

$\times\, d\alpha$, and $\frac{1}{2}v^2 \times gn = \frac{\alpha^2}{2a}$. But the actual celerity is virtually augmented in the ratio of $a$ to $\alpha$; for the pulse of one particle is immediately succeeded by that of the next adjacent, and the tremulous excitement is thus transferred in perpetual succession. Hence $\frac{1}{2}V^2 = gn \times \frac{\alpha^2}{2a} \times \frac{a^2}{\alpha^2} = \frac{1}{2}gna$, and consequently the velocity of internal oscillation is equal to what would be acquired by a body in falling through half the height of the equiponderant column. The velocity with which air would rush into a vacuum is, therefore, $= 8\sqrt{28{,}000} = 1350$ feet *per* second nearly, and the velocity of sound is, according to theory, $= 8\sqrt{14{,}000} = 954$.

## Note XXIX. p. 215.

The theory of undulations, perhaps the most difficult part of physics, was first explained by Newton; and in solving, or at least approximating to the solution of the problem, he has displayed the same wonderful sagacity which, with but feeble aid from the higher calculus, so often conducted him to the noblest discoveries. That great man sometimes employed most judiciously the mere tentative methods, and always shaped his reasonings by a cautious reference to facts and observations. In analysing the very abstruse mechanism of waves, he has indeed committed paralogisms; but these paralogisms are so happily balanced, as not to affect the conclusion. Nor has Lagrange himself, with all his genius and consummate skill, been able to correct in any degree the practical results; for though he has legitimately reduced the conditions of the question to a differential equation, yet this unfortunately is too complex to admit of integration. It has become necessary to restrict the hypothesis, and to reject the higher and apparently less significant terms, before the final value is obtained.

The

The example of the illustrious Newton still deserves our imitation. In many cases, I am persuaded, it would be preferable to make a more sober use of calculation, and to infuse a larger portion of physical principles. Instead of attempting the rigorous solution of a problem, we shall often attain more accuracy and even elegance, by successive approximations. In analysis itself, the direct resolution of equations has been found impracticable beyond the fifth degree; but the humble method by approximation is always simple and expeditious, and perhaps might be preferred even in cubics.

The velocity of sound as deduced from theory is yet considerably different from that which observation assigns. To account for this perplexing discrepancy, various hypotheses have been proposed; but they seem all to be gratuitous, strained, and untenable. It must evidently be referred to some omission or inaccuracy which affects the investigation itself. Perhaps the chief source of error consists in assuming that each contraction is completed before the next begins, or that the elementary pulses form a series strictly of consecutive acts. But while the first molecule obeys its excitement, the impression is partially felt by the second, and thence communicated gradually to the third or even the fourth. Therefore each successive motion is produced by a force not regularly decreasing, but partaking of the uniformity which obtains in projection. Hence the velocity of sound is intermediate between that derived from theory, and that with which the air would rush into a vacuum. But the arithmetical mean of 954 and 1350 is 1152, and the geometrical one is 1135; neither of which differs sensibly from 1142, the quantity determined by actual experiment.

The theory of superficial waves appears still more arduous. Newton has conceived their motions to resemble the oscillations of a liquid contained in a reversed syphon; but this analogy is not more hypothetical, than it is fallacious. And what

what shall we think of the conclusion of Lagrange, that, while the ocean heaves its billows, it is yet seldom agitàted to the depth of an inch? Too many circumstances which affect materially the result, have assuredly been overlooked in the consideration of the problem. A wave does not emerge and then sink down, in the same identical spot: it seems to travel onwards; and this appearance is produced, by the forepart continually rising, while the hinder part of the wave as regularly subsides. The progressive or rather successive transit of the protuberant mass is the necessary result of the reciprocating evagations of its integrant molecules. The fluid particles do not rise and fall perpendicularly; they are supplied and again withdrawn by a lateral motion. The portion directly in front of a wave, feeling the incumbent pressure, rushes forwards, but, suffering continual retardation, it gradually relaxes and accumulates, till, its impetus being spent, it slides back with a gentle descent. Thus each affected particle oscillates in a curve, perhaps the arc of a cycloid. It sets out from below the centre of the wave, and being still urged along, it proceeds with accelerated force until it reaches the anterior margin, and then bending upward with decreasing vigour, it gains the summit, where it is for a moment stationary; the same motion is now repeated, but in a reversed order, and forms the posterior side of the wave.—The figure of the undulations must constitute an important element in the resolution of the problem. The depth of excitement is evidently various, and will depend on the degree and continuance of external agitation. After a violent storm, though the waves have subsided, a swell continues; which proves the existence of short abrupt reciprocations affecting a stratum of very considerable thickness.

## Note XXX. p. 226.

It is plain that a series of equilateral triangles may be continually grouped together, to cover completely any extent of

 surface.

surface. Each point becomes a distinct centre, from which issue lines radiating at angles of 60 degrees. But equilateral pyramids or tetrahedrons can never be piled, to compose a perfect solid. They would leave numerous and wide interstices; and consequently the ranges of divergent points would form winding irregular lines.

## Note XXXI. p. 299.

The term *inertia* denoted originally the natural inaptitude or inherent reluctance to motion, which the ancients, deceived by vague appearances, erroneously attributed to brute terrestrial matter. The famous Kepler, embracing the same notion, but justly considering such tendency to rest as implying a continual effort, was the first who substituted the more apposite expression, *vis inertiæ*. That extraordinary man, whose ardent, penetrating, though irregular genius, armed with unconquerable perseverance, discovered the true laws that regulate the celestial motions, was not equally fortunate in his attempts at their mechanical explication. He supposed that the sun, the centre of the system, formed of "æthereal mould" and animated by a divine or mental energy, diffuses his influence around him in radiating emanations;—that, turning on his axis, this grand luminary involves the planets in the same general revolution;—but that, the primary or *prehensile* force, being modified by the *vis inertiæ* or *renitency* which they severally exert, those sluggish bodies have their periodic motions thence proportionally retarded.

This curious and fanciful hypothesis affords a striking evidence of the slow and imperfect diffusion of knowledge in former times through Europe. Twenty years were elapsed since Galileo had ascertained the real nature of that *inertness* which belongs to matter, and had thereby laid the foundation of genuine dynamics. Yet Kepler was learned, possessed insatiable curiosity, maintained an extensive literary correspondence,

ence, and had repeatedly traversed the spacious theatre of Germany. The Tuscan philosopher proved, that bodies are absolutely passive,—indifferent to motion or rest;—and, if not deranged by the operation of external causes, would either continue in the same position, or would hold a rectilineal and uniform flight. This principle was beautifully brought out and applied in his Dialogues. It is aptly enough expressed by the word *inertia*; but the compound phrase, *vis inertiæ*, should be rejected, as conveying a very different, if not an opposite, idea. Yet the language of Kepler is still used by the bulk of popular writers; and nothing can better show the inveteracy of habit and the power of early prejudice.

## Note XXXII. p. 324.

The expression $dt = \frac{-dh}{ah + h^2}$ is easily integrated, by resolving the fraction into its simple factors. For $\frac{-a}{ah + h^2} = \frac{1}{a+h} - \frac{1}{h}$, and $a\,dt = \frac{dh}{a+h} - \frac{dh}{h}$, whence $at = \text{Log.}\ (a+h) - \text{Log.}\ h + \text{Const.}$ Therefore, when $h = \text{H}$, the constant quantity is $= -\text{Log.}\ (a + \text{H}) + \text{Log.}\ \text{H}$, and applying this correction, we obtain $t = \frac{1}{a}\left(\frac{\text{Log. H}}{\text{Log.}\ h} - \frac{\text{Log.}\ a + \text{H}}{\text{Log.}\ a + h}\right)$.

## Note XXXIII. p. 337.

The measures here assumed differ something from those stated in a former part of the work, because they mark the relative and not the absolute quantities of reflected heat. If unit represents the total reflection from a bright metallic surface, and $\theta$ denotes in millionth parts of an inch the thickness



of

of a film of isinglass superinduced; then will $\frac{100}{100 + \theta}$ in general express very nearly the corresponding modified reflection. Put $110 - 60 \times \frac{100}{100 + \theta} = a$, and, for a six inch ball so coated the time of cooling in minutes is $= \frac{50}{a} \left(\text{Log.} \frac{H}{h} - \text{Log.} \frac{a + H}{a + h}\right)$; where the first three figures are accounted integers.

## Note XXXV. p. 342.

I shall produce a small specimen by way of illustration:—I had a ball of glass blown extremely thin, about $2\frac{1}{4}$ inches in diameter, and with a short narrow neck which admitted a fine thermometer. Being filled with the liquid to be examined, it was warmed by plunging it for a few minutes in hot water; then wiped and planted on a slender inverted tripod, the time was carefully observed of its descent from 30 to 10 degrees above the temperature of the apartment. The results were these:—water, 70 minutes; nitric acid, 56; sulphuric acid, 44; alcohol, 40; and olive oil, 32. Therefore the proportional quantities of heat contained in equal *bulks* of those fluids are as follow:—

| | |
|---|---|
| Water - - - | 100 |
| Nitric acid - - | 80 |
| Sulphuric acid - - | 63 |
| Alcohol - - - | 57 |
| Olive oil - - - | 46 |

These numbers, divided by the corresponding densities, will give the *capacities*, or the measures of heat contained in equal *weights*. But the densities of nitric acid, sulphuric acid, alcohol, and olive oil, are respectively 1.29, 1.85, .89 and .92. Whence

Whence their *capacities*, or specific attractions for heat, range after a different order, thus—

| | | | | |
|---|---|---|---|---|
| Water | - | - | - | 100 |
| Alcohol | - | - | - | 64 |
| Nitric acid | - | | - | 62 |
| Olive oil | - | - | - | 0 |
| Sulphuric acid | - | | - | 34 |

These results in general agree remarkably well with such as have been obtained by other more elaborate methods. Olive oil appears to divaricate the most, for Mr. Kirwan computes its capacity at 71. But the same learned chemist reckons that of linseed oil at only 52; and it seems improbable that two fixed vegetable oils should differ so widely. I presume that some mistake had crept into his first estimate.

## Note XXXV. p. 349.

Let $a$ represent the linear dimension of a solid, whose capacity must therefore be denoted by $a^3$. The differentials are $da$ and $3a^2da$; but $3a^2da = a^3 . 3.\frac{d}{a}$, or is comparatively tripled. In fact, the cube of $1 + \frac{1}{1200}$ may be considered as $1 + \frac{1}{400}$, for when completely expanded it is $1 + \frac{3}{1200} + \frac{3}{(1200)^2} + \frac{1}{(1200)^3}$, of which the third and fourth terms can be safely rejected.

## Note XXXVI. p. 351.

To resolve the differential equation $-dh = hdt\left(\frac{1250+h}{1250}\right)$:

By division, $dt = \frac{-dh}{h}\left(\frac{1250}{1250+h}\right)$, and consequently



$dt =$

$dt = \left(\frac{1250 + h^2}{1250 + h^2}\right) \frac{-dh}{h} - \left(\frac{h^2}{1250 + h^2}\right) \frac{-dh}{h} = \frac{-dh}{h}$ $+ \frac{h\,dh}{1250 + h^2}$. Integrating, therefore, the separate members, we derive $t = -$ Log. $h + \frac{1}{2}$ Log. $(1250 + h^2) +$ Const. Hence the complete integral is $t =$ Log. $\frac{\mathrm{H}}{h} - \frac{1}{2}$ Log. $\frac{1250 + \mathrm{H}^2}{1250 + h^2}$.

## Note XXXVII. p. 352.

The expression $dt = -\frac{dh}{h-h'}\left(\frac{1250}{1250 + h^2 - h'^2}\right)$ is integrated by successive resolution. It is plain that $dt = -\frac{dh}{h-h'}$ $+ \frac{dh}{h-h'}\left(\frac{h^2 - h'^2}{1250 + h^2 - h'^2}\right) = -\frac{dh}{h-h'} + \frac{dh\,(h + h')}{1250 + h^2 - h'^2}$; or, by substitution, $dt = -\frac{dh}{h-h'} + \frac{h\,dh}{a^2 + h^2} + \frac{h'\,dh}{a^2 + h^2}$. Integrating these several terms, therefore, we obtain $t =$ H Log. $(h-h') + \frac{1}{2}$ H Log. $(a^2 + h^2) + \frac{h'}{a}$ arc tang. $\frac{h}{a}$ + Const.—When $a^2$ becomes negative, this mode of integration fails; but, putting $h'^2 - 1250 = \alpha^2$ we have, as before, $dt =$ $-\frac{dh}{h-h'} + \frac{h\,dh}{h^2-\alpha^2} + \frac{h'\,dh}{h^2-\alpha^2}$. This last member is again resolved into $\frac{h'}{2\,\alpha}\left(\frac{dh}{h-\alpha} - \frac{dh}{h+\alpha}\right)$. Hence collecting all the separate parts, we derive by integration, $t = -$ H Log. $(h-h') + \frac{1}{2}$ H Log. $(h^2-\alpha^2) + \frac{h'}{2\,\alpha}\Big($H Log. $(h-\alpha) -$ H Log. $(h+\alpha)$ + Const.

---

The

The scientific reader will perceive, that I have uniformly employed the method of notation which prevails on the continent. In countenancing this innovation, I have not been actuated by an undue predilection for what is foreign, but have adopted it after some experience and mature reflection. Though we dispute with Germany the priority of invention, it must be confessed that the higher calculus was never much cultivated in England. For more than half a century back, we have allowed it to remain almost stationary. And what a vast interval in the career of discovery between Cotes and Lagrange! The mathematicians of the continent have indeed left us so far behind them, that their language and symbols appear at first scarcely intelligible. Our system of notation has remained in the same imperfect state; and unfortunately its narrow basis precludes the possibility of any material improvement. After some hesitation, we have almost universally adopted the nomenclature, though yet imperfect, which is engrafted upon the pneumatic chemistry. And why should we scruple any longer to embrace the consistent and extensive notation appropriated on the continent to the higher calculus? There are manifest indications that this most important study is likely to be revived among us and prosecuted with ardour. And guided by the correct taste derived from our acquaintance with the ancients, we may hope to transfuse into the vast structure of analysis that elegance and luminous connexion which the philosophers abroad have but too much disregarded.

## Note XXXVIII. p. 361.

At first sight it seems a very formidable objection, that ice, being specifically lighter than water, would rise to the surface as fast as it is formed. But this ascensional effort can only be exerted when the congealed mass is surrounded on every side by water. The crystals, shooting from all the prominent points of the bottom, would, by their intertexture, become firmly infixed to the inequalities of the ground and prevent the water from insinuating itself beneath the icy shelve. The continual deposits of sand and mud must likewise contribute to keep it sunk.

I do not mean, however, to assert, that the bottoms of deep lakes are actually filled with perpetual ice; it is sufficient for my purpose, to show that the fact is not without some degree of probability. My arguments are confirmed by some curious and authentic observations made in Siberia. In that remote region, many of the rivers are found to have their beds lined, during the greater part of the year, with a thick crust of ice.

## Note XXXIX. p. 402.

A late ingenious experimenter, who, by the perspicuity and useful tendency of his writings, is deservedly a favourite of the public, has advanced the paradoxical conclusion, that "fluids are non-conductors of heat;" and this strange assertion, from the celebrity of its author, has been treated certainly with more attention and respect than it otherwise merited. If nothing more is meant than, that fluids, as the consequence of their extreme mobility, convey the impressions of heat chiefly by means of their internal motions,—the fact will not be disputed, but, though perhaps more distinctly announced, it can have very little claim to originality. If the proposition, however, be taken in its strict sense, it is most palpably erroneous. Were fluids absolutely incapable of conducting heat, how could they ever become heated? Must we suppose that the particles of a fluid can imbibe heat from those of a solid, and yet not receive it from each other? On mixing cold with hot water, a sort of heterogeneous compound would be formed, each molecule retaining, without participation,

participation, its initial and peculiar temperature. And where no such intermixture can take place, how could water, for instance, be heated by the contact of warm air?—But the question really deserves no serious discussion.

## Note XL. p. 411.

Very few, I believe, have a just conception of the difficulty with which quicksilver moves through narrow tubes. Its progress being always modified by the curvature of its terminating surface, it advances only by successive starts. The first efforts are spent in heaving the central portion; the mercurial column then springs forward into a new position, where it remains till its summit has again swelled to the limit of protrusion.—In attempting some experiments with a sort of manometer whose stem had a bore of about the thirtieth of an inch in diameter, I found the quicksilver, so far from moving uniformly, would, at each step, shoot over a space of almost two inches. But the expansive power which heat communicates to that liquid metal surpasses prodigiously the mere elastic force of air. Hence the nice regularity of mercurial thermometers, notwithstanding the fineness of their tubes. The vast difference, however, between the propulsive energies of heated air and heated mercury, will appear more striking from a simple calculation. Mercury, I find, scarcely suffers a contraction corresponding to the 400th part of a degree, when subjected to the additional pressure of a whole atmosphere. But with one degree of heat, air expands the 250th part of its bulk. Therefore 400 × 250, or 100,000, must express the superior expansive energy of the thermometer in comparison with the manometer; a superiority of force so vast as to be capable of overcoming every species of obstruction.

## Note XLI. p. 416.

The hydrogenous gas introduced into the balls was not indeed absolutely free from an admixture of atmospheric air; yet, considering the precautions that were used, the residuum which still adhered must have been extremely minute, and quite insufficient, I presume, to produce such a notable effect. I am therefore inclined to suspect that even the purest hydrogenous gas has always combined with it a certain small proportion of oxygen. That gas is obtained only from the decomposition of water, by the attractive energy of zinc or iron, whether assisted by the dilating power of heat or the conspiring affinity of sulphuric acid. Every chemical arrangement supposes, however, a mutual compensation of forces; and consequently the whole of the oxygen is not absorbed by the metal, but some part of it is seized and transported by the hydrogenous gas.

Nor is it difficult to explain the effects recited in the text. The portion of oxygen thus concealed in hydrogenous gas is slowly detached, by the attraction of the deliquiate potash; and being communicated to the carmine, dissolves it and makes it to precipitate in flocules. But sulphuric acid is already saturated with oxygen, and therefore has no disposition to absorb that element.

## Note XLII. p. 437.

The measures that have been assigned for the photometer are not merely the result of trial, but partly founded on a simple calculation, which I shall here subjoin.—The dimension of either ball and the width of the bore of a differential thermometer being given, it is easy to compute the size of the scale. This problem comprizes two cases.—1. When the tube has a horizontal position.—Let A denote the diameter of either ball, and $a$ that of the bore of the tube. Then if $x$ denotes

notes the space through which the coloured liquor has moved, $\frac{\pi}{4} \cdot a^2 x$ must represent the corresponding cylinder of air thus displaced. But the capacity of either ball is $\frac{\pi}{6} \cdot A^3$; and therefore the air contained within the one will have its density increased, and that within the other will have it diminished, by the portion $\frac{\frac{\pi}{4} a^2 x}{\frac{\pi}{6} A^3}$, or $\frac{3 a^2 x}{2 A^3}$. The difference between the opposite forces is consequently $\frac{3 a^2 x}{A^3}$. But since each degree of heat augments the elasticity of air by the 250th part, we shall have for the extent of 100 photometric degrees the equation $\frac{1}{25} = \frac{3 a^2 x}{A^3}$. Hence $A^3 = 75 a^2 x$, and $x = \frac{A^3}{75 a^2}$.

2. When the graduated branch has a vertical position.—This is the most ordinary construction, and the only one which I have employed for the photometer.—To the difference already stated between the opposite elastic forces, there is now added the weight of the column raised. But this accession of force is denoted by $\frac{x}{225}$, since a column of sulphuric acid of about 225 inches in height may be computed as equivalent to the pressnre of the atmosphere, or $29.8 \times \frac{13.6}{1.8} = 225$. Therefore $\frac{1}{25} = \frac{3 a^2 x}{A^3} + \frac{x}{225}$, and by reduction $9 = \frac{675 a^2 x}{A^3} + x$, or $9 A^3 = 675 a^2 x + A^3 x$, and consequently $x = \frac{9 A^3}{A^3 + 675 a^2}$ But 676 being the square of

of 26, the value of $x$ may be expressed, more simply and almost as exactly, by $\frac{9A^3}{A^3 + (26a)^2}$.—Thus, for example, if $A = .5$, $a = \frac{1}{52}$; then $x = \frac{9A^3}{A^3 + (.5)^2} = \frac{1.125}{.125 + .25} = 3$.

There is evidently a limit to the magnitude of the scale, however much the bore may be contracted or the balls enlarged; for $x$ is ultimately $= \frac{9A^3}{A^3 + 0}$, or 9.

## Note XLIII. p. 453.

Sometime after I had formed this opinion, I felt the most agreeable surprize to find, on consulting Riccioli's *Almagestum Novum*, that the same idea was entertained by more than one astronomer, about the middle of the seventeenth century. I know not why a conjecture so specious has been suffered to fall into oblivion. The moon, examined with a good telescope, presents an appearance altogether different from the aspect of our globe. Those sharp ledges of rock, and those singular cavities or knolls, exactly circular and often concentric, so frequent on her disc, bear no resemblance to any thing that is seen upon earth. The most fanciful names have been imposed, but which convey no distinct images of the objects signified. No forests or lakes appear in the moon; no cultivation, and no cities, the abodes of any tribes of gregarious animals; the only hemisphere ever turned to our view exhibits still the same unvaried picture of silence, solitude, and gloom. We seem to contemplate only a white incinerated mass. Some round spots, of dazzling brightness, are like so many fountains of light; while wide-extended plains, their phosphorescent virtue nearly spent, already look obscure, and verge towards extinction. Reasoning from analogy, therefore, we should expect the lunar surface to grow continually paler, till it becomes quite effete and overspread with

with a dusky shade. It may only then be fitted for the reception of something akin to the humbler plants and the simple rudiments of animal existence. Yet the progress of obscuration must be wonderfully slow. The most ancient writers were acquainted with spots on the moon's disc, and considered them as earthy impurities absorbed amid her æthereal or aqueous nutriment.

The theory which I have now stated, will explain the familiar appearance of "the old moon in the new moon's lap." After emerging from conjunction with the sun, her sharp horns are seen connected by a silver thread or lucid bow, which completes the circle; and a very faint light seems to be suffused over the included space. This bright arch, however, becomes always less vivid, and before the moon is 5 or 6 days old, it has almost totally vanished.—The pale outline of the old moon is commonly ascribed to the reflection or secondary illumination from the earth. But if it were derived from that source, it would appear densest near the centre, and gradually more dilute towards the edge. I rather should refer it to the spontaneous light which the moon may continue to emit for some time after her phosphorescent substance has been excited by the action of the solar beams. The lunar disc is visible, although completely covered by the shadow of the earth; nor can this fact be explained by the inflection of the sun's rays in passing through our atmosphere, for why does the rim appear so brilliant? Any such inflection could only produce a diffuse light, obscurely tinging the boundaries of the lunar orb. And in this case, the earth, presenting its dark side to the moon, would have no power to heighten the effect by reflection. But even when this reflection is greatest about the time of conjunction, its influence seems extremely feeble. The lucid bounding arc is occasioned by the narrow *lunula*, which, having recently felt the solar impression, still continues to shine, and, from its extreme obliquity, glows with concentrated effect.

It

It is possible to imitate the lunar surface, with all its irregular distribution of light and shade, by a very simple experiment. Introduce a bit of phosphorus into a glass ball of two or three inches in diameter, and having heated it to catch fire, keep turning the ball round, till half the inner surface being covered with melted phosphorus, the inflammation has ceased. There is left a whitish crust or lining, which, in a dark place, will shine for some considerable time. Broad spaces will assume by degrees an obscure aspect, while circular spots, frequently interspersed, will yet glow with a vivid lustre.

## Note XLIV. p. 455.

No person, I believe, on first viewing the spectrum, would, by a spontaneous movement, pronounce it to contain seven distinct colours. He would most probably, if not biassed by some preconceived notions, fix on a smaller number of radical divisions, with various intermediate shades. In fact, every possible tint may be produced by the due mixture of only three ingredients—red, yellow, and blue.

The capital experiment from which Newton inferred the constant proportionality of the coloured spaces in the spectrum, has been found to be fallacious and inconclusive. The zealous attempts of our countryman Dolland to defend that proposition, while they totally failed of success, fortunately led him to the valuable invention of the achromatic telescope. The dispersive quality of a prism is not deducible from its mean refractive power. No common affection pervades all the rays, but each of them suffers an effect peculiar to itself, and which seems to depend merely on its specific relation to the diaphanous medium. The construction of achromatic glasses plainly supposes such a principle. Yet, with what tardy reluctance does mankind admit the consequences of any discovery! Though half a century has passed away since

since that useful improvement was made, the earlier opinions still continue to be repeated in the elementary books, nor is their incongruity distinctly perceived even by authors of note. The whimsical project of an *ocular harpsichord* will long attract the gaze of vulgar admiration.

## Note XLV. p. 458.

A celebrated and most successful astronomical observer, who sometimes indulges a great latitude of fancy, and espouses opinions that are hardly consistent with the sober pretensions of science, has drawn inferences of a very curious nature from certain experiments made on the several calorific energies of the prismatic colours. He advances, that the rays of the sun consist of two distinct kinds, the one being endued with the exclusive property of illumination, and the other having only the power of giving heat;—that they are both of them susceptible of reflection and refraction;—and that, though generally blended together, the latter sort is somewhat less refrangible.—This hypothesis, how much soever discordant with the general phænomena, has been treated with that partiality and timid deference, which the glare of paradox and the weight of authority seldom fail to produce on the bulk of men. It is a memorable instance of the fallacy of detached observations, when not directed with circumspection, and not interpreted by the concurring lights of extended analogy. Most of the collateral experiments, and which relate to culinary heat, are contradicted by the whole tenor of the present Inquiry; and I am persuaded that if the fundamental fact were examined fairly with sedulous attention, it would be likewise found to originate in mistake. I will not seek to trace the various sources of error, but I shall mention a single circumstance which we are most apt to overlook, and which might yet occasion egregious deception.

A prism fixed in a slit of the window-shutter, must evidently,

dently, besides the direct beams of the sun, admit the light sent from the adjacent portion of the sky, and which, in this climate, is always very considerable. These extraneous rays, being intermixed in refraction, will form a white space, of various degrees of intensity, and extending on both sides of the coloured spectrum, but chiefly at the boundary of the red. In certain positions of the prism, this effect becomes concentrated; for the quantity of refraction which it occasions is not constant; but increases considerably with the obliquity either of incidence or emergence. On presenting an equilateral prism of flint glass, with its base upwards, to the sun, the slanting rays which first enter it suffer the greatest refraction. Turning it about its axis, the spectrum will gradually descend with almost an equal measure of angular motion; and consequently the light which now comes to be admitted from the bright expanse above the sun, is refracted to nearly the same elevation, and thus collected immediately below the spectrum. Continuing to turn the prism, the quantity of refraction will still diminish, but always more slowly, till it is contracted to its *minimum*, when the incident and the mean emergent beam forms each an angle of 53° 8′. The extraneous rays have here their natural divergence, being neither condensed nor dilated. Turning the prism still in the same direction, the spectrum, after keeping for a moment stationary, will again ascend with increasing rapidity and extension, till the angle of incidence becomes about 36°, when it will finally vanish. At this limit, the adventitious light is extremely diffuse; but the colours of the spectrum itself are likewise much attenuated. For ascertaining their calorific powers, the best position of the prism is between this and the stationary limit.

An example will elucidate these remarks. Suppose the angle of incidence to be 80°; the corresponding refraction is then 56° 51′. But the quantity of refraction due to an incidence of 85° is 60° 54′, or 4° 3′ more; and consequently the indirect

indirect light received from a space of 5° above the sun is concentrated into a spot of 57′ under the spectrum, whose breadth is about 3°.—The refraction corresponding to 75° of incidence is only 53° 22′, or 3° 29′ less than that of the mean emergent ray. Wherefore the light sent from the portion of the sky 5° below the sun, occupying after reflection a breadth of 1° 31′, is confounded with the spectrum itself. But towards the evanescent limit, the relations are materially changed. Thus the angle of incidence being 40°, that of refraction is 51° 22′; and when the former is 45°, the latter is 47° 47′: consequently the light received from the space of 5° above the sun, is spread into 8° 35′. Taking the whole compass of the prism, however, there is a predominant disposition to concentrate the extraneous light, and hence the chance lies evidently on the side of error.

---

Before I close this volume, I may notice a simple improvement or modification of the differential thermometer, which fits it for estimating with nice precision the intensity of the diffuse radiations of heat. It has still the form represented by fig. 2, only the ball of the graduated stem is completely gilt or enamelled with gold. But the two balls, exposed to the same influence, will now receive very different impressions, and the excess of energy, which the instrument marks, must therefore amount nearly to seven-eighths of the whole vibratory tide. Hence it will measure the quantities of heat that are continually thrown from the fire into a room. We can thus calculate, with equal ease and certainty, the relative advantages arising from various constructions of chimneys.

I shall conclude with the explication of a curious experiment related by Mr. T. Wedgwood, in his excellent paper inserted in the Philosophical Transactions for the year 1792. With a view to discover how far the presence of light was capable of augmenting the effect of heat, that ingenious inquirer cemented two small cylinders of polished silver, one of which was covered externally with a thin coat of blacking, into the bottom of an earthen tube, and held it over burning charcoal contained in a crucible. Accordingly, looking down the tube, the end of the blackened cylinder appeared to grow much sooner red than the other; but on removing it from the charcoal fire, the same cylinder likewise ceased much sooner to shine.—These are manifestly the distinguishing effects of radiant heat, which, from a painted surface of any colour, is the most copiously both absorbed and discharged. The light emitted from the burning charcoal had comparatively a very slight influence to accelerate the main impression.

FINIS.

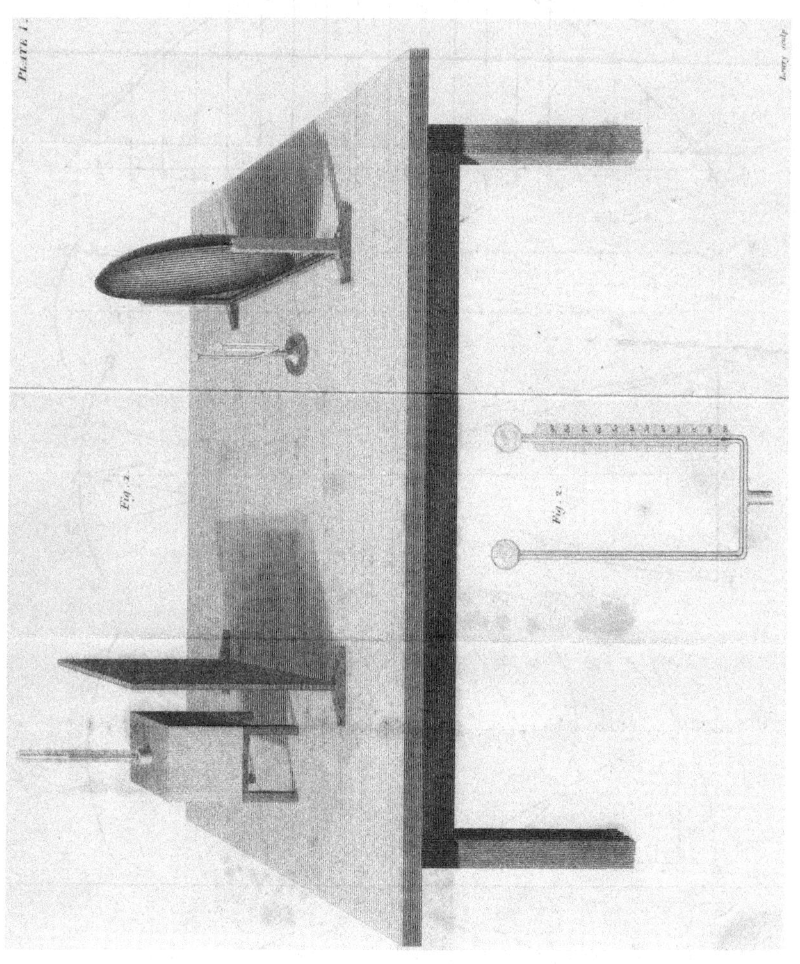


The material originally positioned here is too large for reproduction in this reissue. A PDF can be downloaded from the web address given on page iv of this book, by clicking on 'Resources Available'.

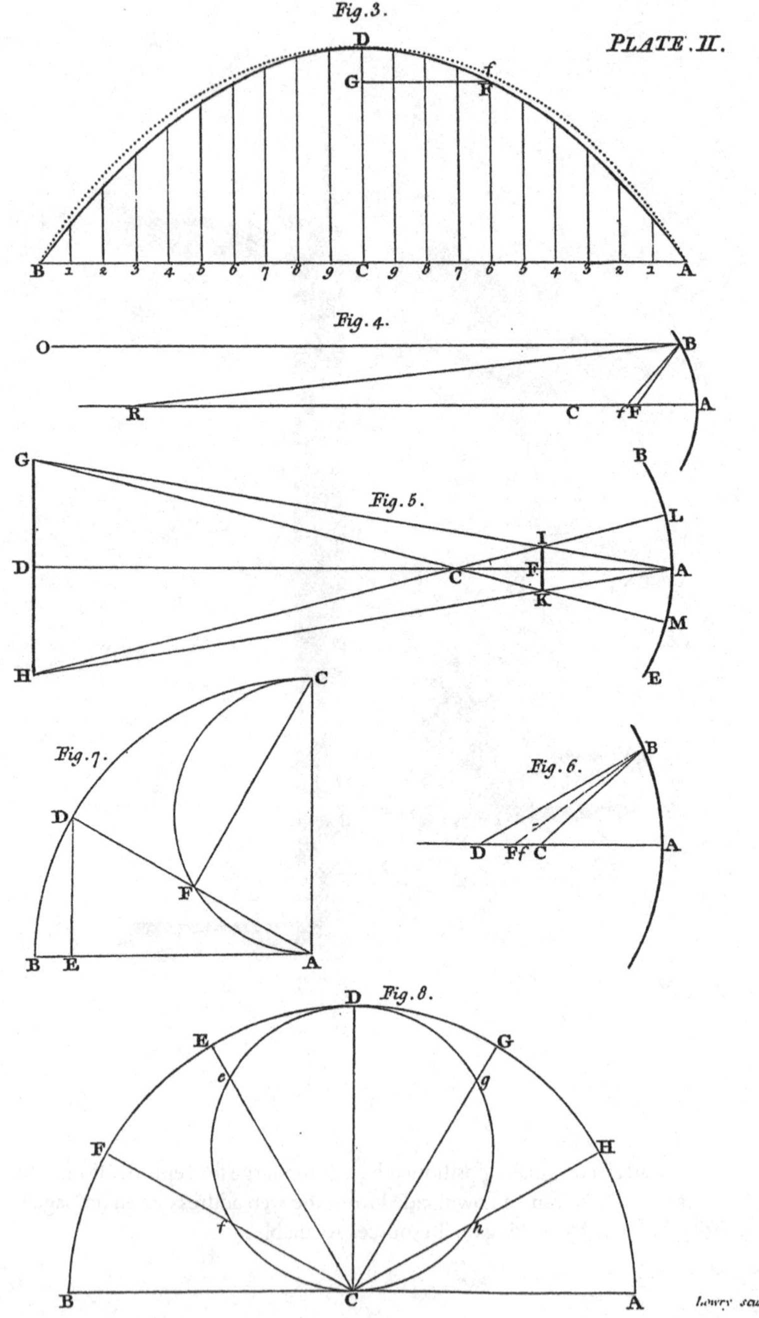
PLATE . II.
Fig. 3.
D
f
G
F
B
1
2
3
4
5
6
7
8
9
C
9
8
7
6
5
4
3
2
1
A
Fig. 4.
O
B
R
C
f F
A
Fig. 5.
G
B
L
I
D
C
F
A
K
M
H
E
Fig. 7.
C
D
F
B
E
A
Fig. 6.
B
D
F f
C
A
Fig. 8.
D
E
G
e
g
F
H
f
h
B
C
A
Lowry sculp.

PLATE III.

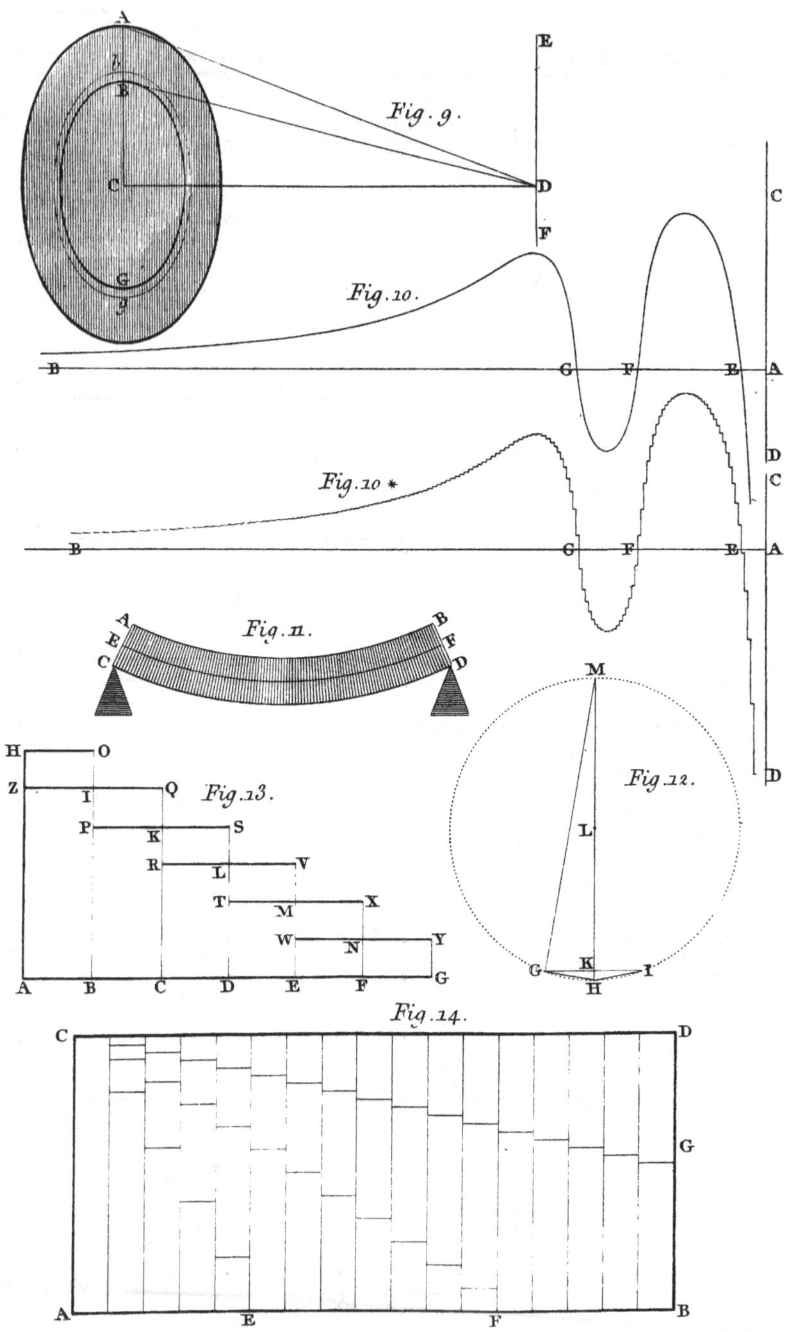


Lowry sculp.

PLATE IV.

Fig. 15.

Fig. 17.

Fig. 16.

Fig. 18.

Fig. 19.

Fig. 20.

Fig. 25.

Fig. 21.

Fig. 24.

Fig. 26.

Fig. 22.

Fig. 23.

Lowry sculp.

PLATE V.

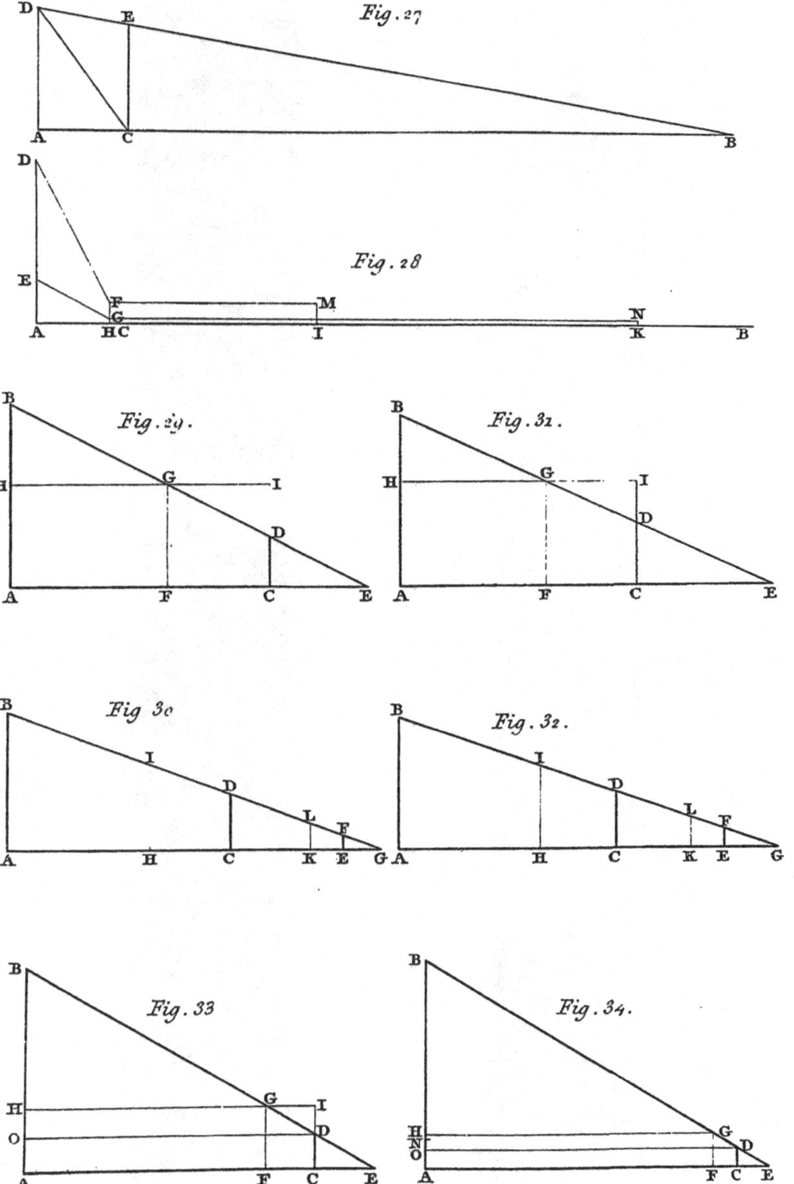


Lowry sculp

PLATE VI.

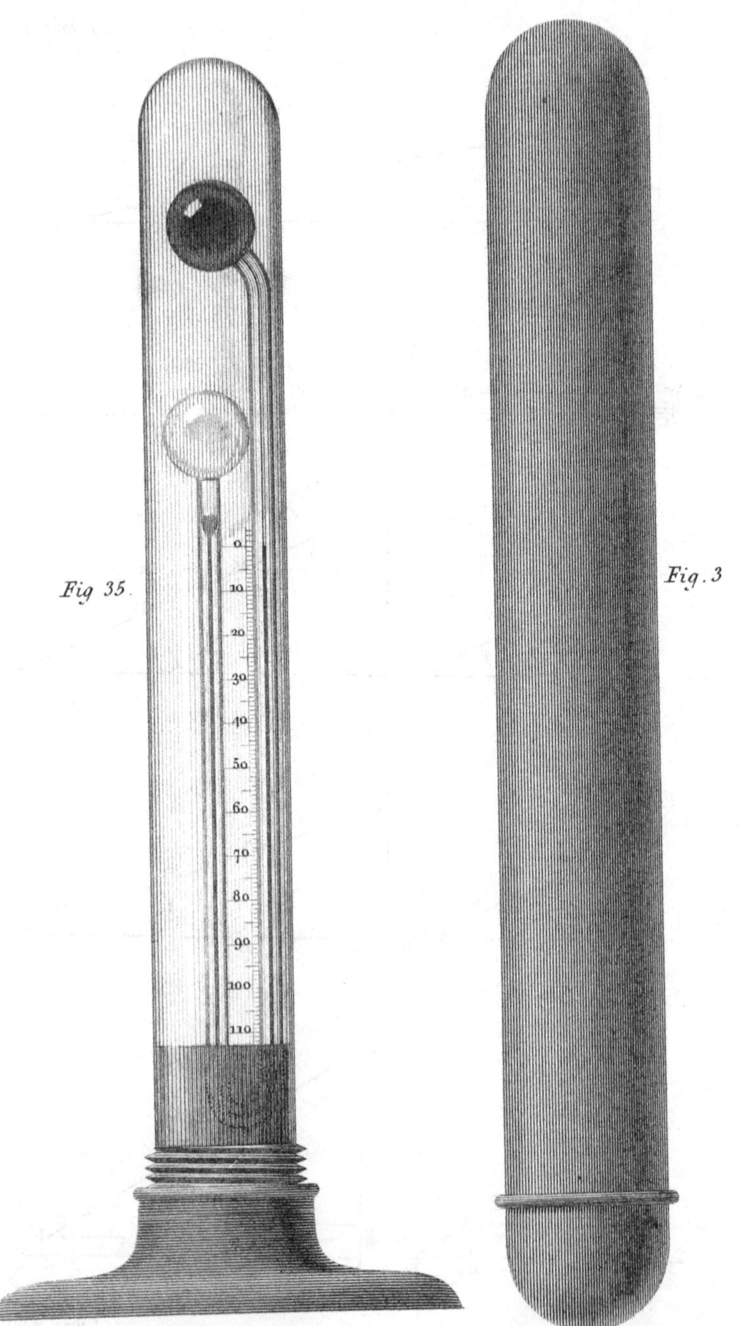


Fig 35.

Fig. 3

Lowry sculp.

PLATE VII.

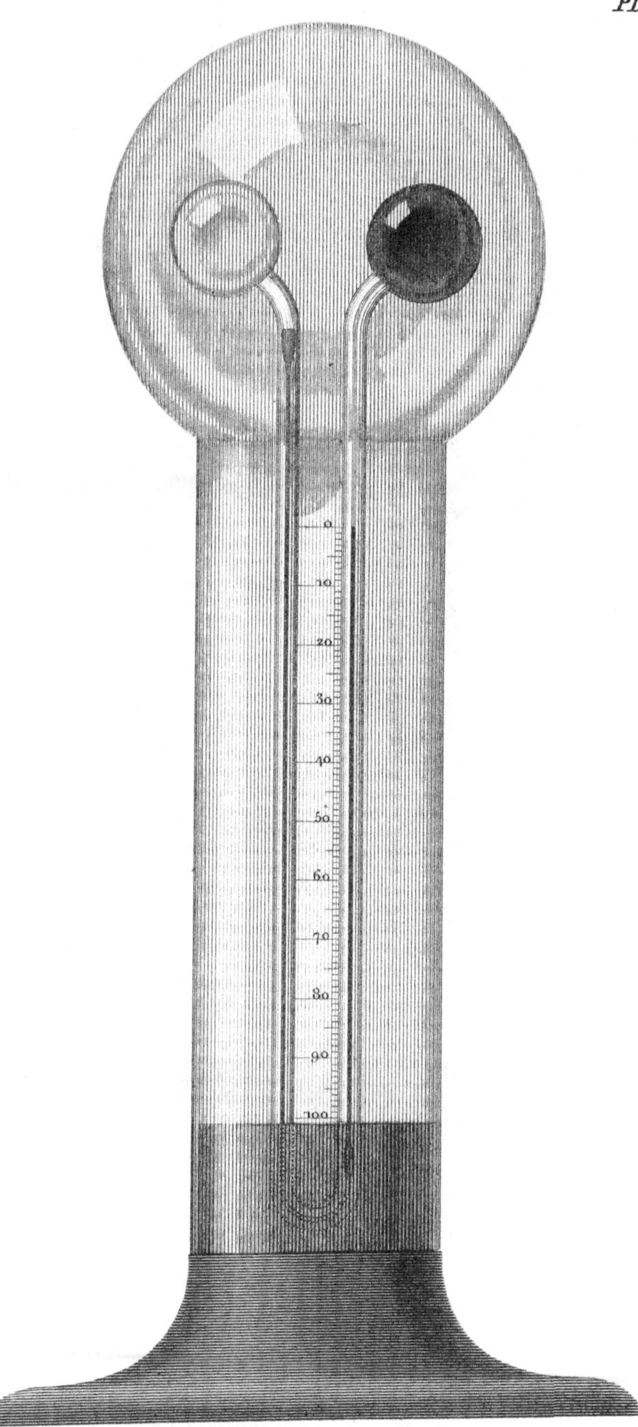


Lowry sculp

PLATE VIII.

Fig. 37

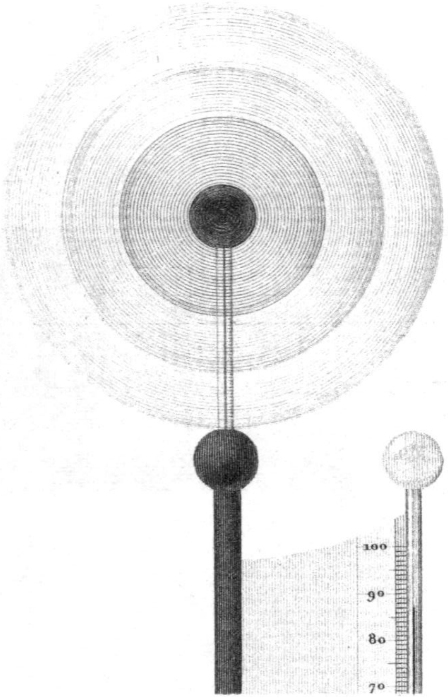

Fig. 38.

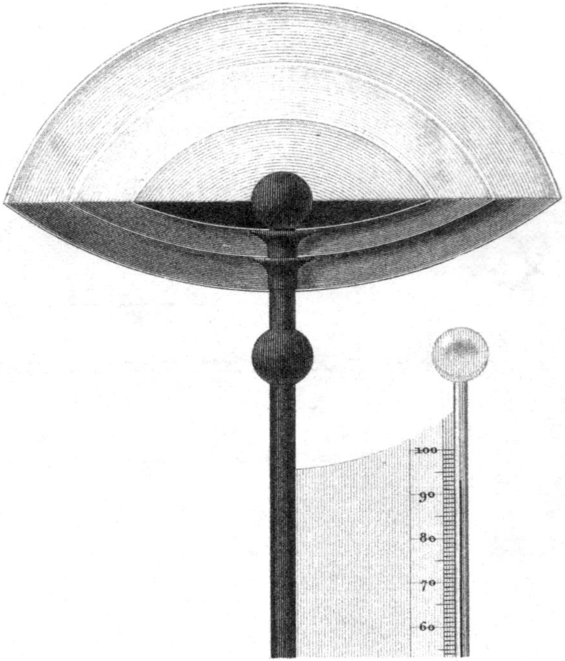

Lowry sculp.

PLATE IX.

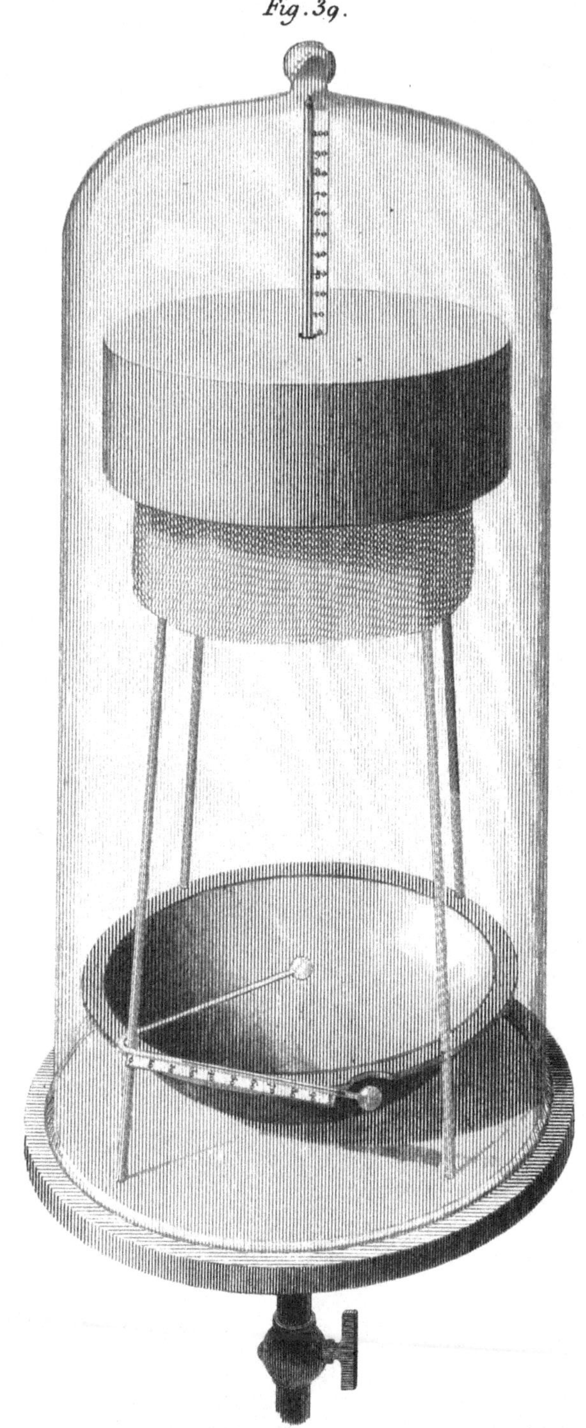

Fig. 39.

Lowry sculp.

Printed in the United States
By Bookmasters